A PLUME BOOK

WHEN SCIENCE GOES WRONG

Simon LeVay is a British-born neuroscientist who has served on the faculties of Harvard Medical School (1971–1984) and the Salk Institute (1984–1992). He is best known for a 1991 study published in *Science* that reported on a difference in brain structure between homosexual and heterosexual men. Since 1992 he has been a writer based in West Hollywood, California. He has written or coauthored eight previous books, including *The Sexual Brain, Queer Science*, and the college textbook *Human Sexuality* (with Sharon Valente). He was the 1963 bicycle hill-climb champion of East Anglia, a perfectly flat region of England.

D1548871

WHEN
SCIENCE
GOES WRONG

Twelve Tales from the Dark Side of Discovery

Simon LeVay

A PLUME BOOK

PLUME
Published by the Penguin Group
Penguin Group (USA) Inc., 375 Hudson Street, New York, New York 10014, U.S.A.
Penguin Group (Canada), 90 Eglinton Avenue East, Suite 700, Toronto, Ontario, Canada M4P
2Y3 (a division of Pearson Penguin Canada Inc.); Penguin Books Ltd., 80 Strand, London
WC2R 0RL, England; Penguin Ireland, 25 St. Stephen's Green, Dublin 2, Ireland (a division of
Penguin Books Ltd.); Penguin Group (Australia), 250 Camberwell Road, Camberwell, Victoria
3124, Australia (a division of Pearson Australia Group Pty. Ltd.); Penguin Books India Pvt.
Ltd., 11 Community Centre, Panchsheel Park, New Delhi – 110 017, India; Penguin Group
(NZ), 67 Apollo Drive, Rosedale, North Shore 0632, New Zealand
(a division of Pearson New Zealand Ltd.); Penguin Books (South Africa) (Pty.) Ltd.,
24 Sturdee Avenue, Rosebank, Johannesburg 2196, South Africa

Penguin Books Ltd., Registered Offices: 80 Strand, London WC2R 0RL, England

First published by Plume, a member of Penguin Group (USA) Inc.

First Printing, April 2008
1 3 5 7 9 10 8 6 4 2

LIBRARY OF CONGRESS CATALOGING-IN-PUBLICATION DATA
LeVay, Simon.
When science goes wrong : twelve tales from the dark side of discovery / Simon LeVay.
p.   cm.
"A Plume book."
ISBN 978-0-452-28932-1 (trade pbk.)
1. Errors, Scientific—Popular works.   2. Science—Miscellanea.   I. Title.
Q172.5.E77L48 2008
500—dc22   2007023389

Printed in the United States of America
Set in Sabon
Designed by Leonard Telesca

Contents

Preface

Mostly, we hear about science's triumphs—the wonder drugs, the moon landings, the ever-faster computers. But for every brilliant scientific success there are a dozen failures. Usually these involve no more than some wasted dollars and a blank spot on someone's résumé. Once in a while, though, science doesn't just fail—it goes spectacularly, even horribly, wrong. And that makes for a great story.

This book is a collection of twelve such stories. They are linked by the common thread of scientific failure, but in other respects they are quite diverse. I wanted, in the first place, to range over many different kinds of science. Some of the chapters relate to explorations within the basic sciences, such as nuclear chemistry, volcanology, and neuroscience. The majority, however, focus on the applied sciences, because it is when science serves human ends that the opportunities for truly memorable screw-ups are most likely to arise. These sciences include medical research, forensic science, meteorology, microbiology, and psychology. One of the stories deals with an engineering failure, but it was a failure rooted in scientific error.

In addition, I wanted to illustrate the rich variety of ways in which the scientific process can go awry. Failures, disasters, and other negative outcomes of science can result not only from bad luck, but also from the failure to follow appropriate procedures or to heed warnings, from confusion of units, from ethical breaches in the treatment of human subjects, from the pressure to get quick results, from excessive ambition or financial greed, from the failure to think broadly enough about the consequences of one's work, or from fraud. Or even from a couple of mislabeled bottles. The stories in this book illustrate the consequences of many of these factors, acting alone or in diabolical combination.

This book is not an attack on science. I am a scientist myself, and I consider science to be one of the most beautiful, challenging, and worthwhile activities that humans can engage in. The events described in this book are no more the story of science than plane crashes are the story of aviation. If I thought that the publication of this book would bring the entire cavalcade of science to a jangling halt—or even impede its progress in the slightest degree—I would not have written it.

The book is also not intended to be a complete or academic survey of scientific failure. There are plenty of sciences that escape mention in these pages, plenty of failure modes that I don't discuss, and plenty of errors more laughable, accidents more tragic, and wrongdoings more egregious. Rather than trying to be comprehensive, I have followed the advice offered to the historian by Lytton Strachey in *Eminent Victorians:* "He will row out over that great ocean of material, and lower down into it, here and there, a little bucket, which will bring up to the light of day some characteristic specimen, from those far depths, to be examined with a careful curiosity."

In keeping with this approach, I have not arranged my "specimens" into any logical order, based for example on their historical sequence, the scientific disciplines they relate to, or the modes of scientific failure that they exemplify. Instead, I have laid them out on the deck in what I hope is an agreeable pattern, juxtaposing light and dark, new and old, innocent and malevolent. If they coalesce into a larger picture, so much the better.

It is customary in a book of this kind to thank one's sources, but I do so with particular sincerity in this case, because some of the people I interviewed were agreeing to talk about episodes in their lives that they would probably rather forget. They had little to gain from reliving those episodes, and I thank them for doing so.

This book is not just about scientific error and wrongdoing, however; it is also about bravery in the face of danger, endurance in the face of suffering and loss, intelligence and persistence in the search for causes, and even sometimes about the *right* way of doing science. Some of my interviewees exemplify those traits, and I thank them sincerely too.

The complete list of the people I interviewed is as follows: Ken Alibek, Ph.D., Nicoline Ambrose, Ph.D., Colin Blakemore, Ph.D., Oliver Bloodstein, Ph.D., Arthur Caplan, Ph.D., John Casani, Ph.D., Bernard Chouet, Ph.D., Rick Doblin, Ph.D., Raymon Durso, M.D., Robert Erickson, M.D., Michael Fish, Rebecca Folkerth, M.D., Curt Freed, M.D., Paul Gelsinger, Bill Giles, Jack Green, Ph.D., Kenneth Gregorich, Ph.D., Charles Grob, M.D., Jeanne Guillemin, Ph.D., Peter Gumbel, Stephen Hanauer, Ph.D., Anita Hart, Robert Iacono, M.D., Steve Jolly, Ph.D., Thomas Jung, Ph.D., Walter Loveland, Ph.D., Ewen McCallum, M.Sc., Matthew Meselson, Ph.D., Victor Ninov, Ph.D., George Ricaurte, M.D., Ph.D., Richard Schwartz, Ph.D., William Thompson, J.D., Ph.D., Sam Thurman, Ph.D., Don L. Truex, D.D.S., Kay Truex De Justo, Inder Verma, Ph.D., and Charles Wood, Ph.D.

There were also people who played key roles in the events described in this book but who declined to speak with me, or who did not respond to repeated requests for an interview. I have attempted to describe their actions and represent their viewpoints as accurately and fairly as possible by reference to their published writings or statements, news reports, or information provided by other interviewees. Any failure to do them justice represents a shortcoming in this book that I regret.

I am grateful to Christian Wehrhahn, Ulrike Seibt, Kerry Sieh, and my brother, Benedict le Vay (author of the *Eccentric Britain* series), for reading and commenting on portions of the book. Ben also helped develop the book's concept, and my agent Andrew Lownie helped turn that concept into a proposal for a saleable manuscript. So all I had to do was write it.

CHAPTER 1

NEUROSCIENCE: The Runner's Brain

Morgues are spooky places at the best of times. Even during the day, when knots of chattering medical students gather round the brightly lit dissecting tables and senior pathologists poke at ruptured aortas or cancer-ravaged livers, morgues offer uncomfortable reminders of our own mortality. Not just the sight of the dead and the disorders that killed them, but also the odor—if not the odor of death itself, then the acrid fumes of formalin, the preservative that keeps death's putrefaction at bay.

At night, it's worse. And it *was* night—the middle of the night—as Rebecca Folkerth stood next to a table in the morgue at the New England Medical Center, Tufts University's teaching hospital on Tremont Street in downtown Boston. It was on a weekend in the late winter of 1991. Bundled against the cold, a few late-night revelers were still strolling the streets outside, looking for all-night Chinese restaurants or checking out the attractions of the Combat Zone, Boston's always-open red-light district. But inside the morgue, all was quiet, and Folkerth was alone.

Well, not quite alone. For company she had fifty-three-year-old Max Truex, whose brain she was removing.

Folkerth was a pathologist, but not a senior one. In fact, the blond thirty-two-year-old had just taken her boards—the examination that qualified her in the subspecialty of neuropathology—a few months earlier. She hadn't yet accumulated the portfolio of macabre experiences that make pathologists such entertaining dinner guests, but this night would get her off to a good start. As she looked at Truex's brain, she blinked, looked again, and said to herself, "What the *hell* is this? This is *creepy*!"

Max Truex was born in Warsaw, Indiana, in 1935. His father, Russell, was a locomotive engineer with the Pennsylvania Railroad. Although he drank too much, Russell was a good family man and a reliable provider for his wife, Lucile, and their three sons, Gene, Max, and Don.

Gene was four years older than Max, so there was always a certain distance, an unquestioned division of authority and obedience, between them. With Don it was different. Don was born only a year and a half after Max. What was worse, Max grew slowly, so already by the time he was six Don had caught up with him in stature. When I visited Don in 2005—he's now in his late sixties, a practicing dentist in Santa Barbara, California, and a running, cycling, and general fitness enthusiast—he told me that the height issue was a major factor in their relationship. "It was a real sore point for him," he said. "When we were six, eight, and ten years old it was a fight every day. Our next-door neighbor said she thought we would never grow up, because one of us was going to kill the other."

But it wasn't either Max or Don who failed to grow up—it was their older brother, Gene. When he was sixteen, and the younger boys were twelve and eleven, the three of them were driving in the family car, with Gene at the wheel. At an intersection they collided with a dump truck, and Gene was killed.

After this tragedy, Max found himself suddenly and unexpectedly in the role of eldest son, yet with a younger brother who was now several inches taller than he was. What saved the two boys from mutual destruction was sports. Gene, before his death, had already been running the mile in high school, and now the two younger boys followed his lead, taking to running and other sports with fierce dedication and competitiveness. And in the process they became good friends.

Although there are exceptions, taller runners usually excel at sprinting, while smaller runners do better in endurance events. At a final height of five feet and five inches and a weight of 130 pounds, Max Truex was very similar in stature to the Ethiopians like Haile Gebrselassie, who utterly dominated world competition at 10,000 meters during the 1990s, or Kenenisa Bekele, the current world record–holder at that distance. But back in 1950, no one had set eyes on an Ethiopian runner, so Max's modest size drew people's attention. It encouraged them to develop an affectionate or protective attitude toward him, as if they saw him as a permanent child.

There were other traits that had the same effect. Although the Truexes are French in ancestry, Max had a broad, Slavic-looking face that seemed to carry a fixed, somewhat childish smile, even when he was running. And as he ran, he "skipped"—he would frequently and erratically switch his stride, as if out of sheer playfulness. Indeed, he was naturally good-humored. For all these reasons, people called him Maxie and liked to take him under their wing. Later, when newspapers started recording his feats, they would refer to him with patronizing titles like "The Little Strider." They described him as "spunky," as if his small stature was a natural handicap that only grit and determination had overcome.

In high school, Max Truex was a star athlete—he won the state cross-country championship and set a national interscholastic record for the mile. On account of his performances, Truex was actively recruited by several universities, and he finally accepted a track scholarship at the University of Southern California, where he was drawn by the warm climate and the school's high ranking in athletics.

One of the people who was involved in recruiting him was USC's assistant track coach, Jim Slosson, who became a lifelong advisor and friend to Truex. Under Slosson's tutelage, Truex quickly came to focus on the long-distance events—the 5,000 and 10,000 meters— where his remarkable powers of endurance counted the most.

Success came quickly. In June of 1956, at the age of twenty, Truex won the 10,000 meters at the U.S. Olympic trials, thus guaranteeing him a berth on the Olympic team. In October of that year, one month before the Games, he set his first U.S. national record, in the 5,000 meters. (He was to set four more records in that event over the following six years.) Unfortunately, he suffered a muscle injury shortly before leaving for the Olympics, which were held in Melbourne, Australia. He competed in the 10K but did not do well.

At USC, Truex was a member of the Air Force Reserve Officers Training Corps, so in 1958, after graduating, he joined the Air Force and served for four years as an officer at Oxnard Air Force Base. This base was a hotbed of athletic activity—the athletes were given all the time they needed to train and compete. In 1959 Truex set a new U.S. indoor record in the 2-mile event. Then, in the following year, he was the highest-placed American in the 10,000 meters race at the U.S. Olympic trials, again guaranteeing himself a place on the Olympic team.

The 10K race at the Rome Olympics, in September of 1960, was probably the high point of Truex's running career. He didn't win—in fact, he only finished sixth—but his time of 28 minutes and 50.34 seconds took eight seconds off the existing U.S. record. Truex's performance put the United States on the distance-running map for the first time. Just a week after the Rome Olympics, Truex iced the cake by setting a new U.S. record in the 3,000 meters.

Truex quit competitive athletics in 1962, when he left the Air Force and entered law school at USC. He didn't quit running, however. He ran all through his three years at law school, and he ran while he was working as an attorney, first in private practice in Orange County, and then in the legal office of the County of Los Angeles. He lived in an apartment near Universal Studios: During that period he ran five, six, or seven miles daily along a footpath next to the Hollywood Freeway. He didn't need the spur of competition, he just loved running and the sense of fitness that went with it.

"I told him I thought it was unhealthy, to run along the freeway," Jim Slosson told me in a 2005 interview, and indeed it must have been. Los Angeles at that time had some of the worst air pollution in the country. Truex was breathing in a truly evil brew of toxins, including carbon monoxide, particulates, ozone, and lead from automobiles on the freeway, as well as industrial chemicals such as the pesticide

DDT, which was being manufactured with carefree abandon by the Montrose Chemical Corporation at its plant on Normandie Avenue. Although now banned, residues of the wind-born DDT dust can still be found in soils miles from the plant.

Even during Truex's competitive career, there were several occasions when his running had caused him acute health problems. He had to drop out of at least three races on account of exhaustion caused by some combination of excessive heat and air pollution. The worst occasion was in 1961, while he was competing in the Corrida de São Silvestre, a traditional 15-kilometer race that is held every New Year's Eve in São Paulo, Brazil. Although the race took place at night, it was oppressively hot and humid. Truex was in the lead, immediately behind a phalanx of motorcycles and television trucks that belched the combustion products of cheap South American gasoline. He suddenly collapsed, and the next thing he knew he was in the hospital, hooked up to an IV but losing fluids faster than they could be pumped in. By the time he got on a plane back to the United States, he was fifteen pounds lighter than when he set out.

In spite of his chronic exposure to pollutants and heat stress, Truex remained in apparent good health during his early professional years. He enjoyed his work as an attorney, often appearing in court to argue real-estate and land-use cases on behalf of the county. And in 1973 he married.

I recently met Truex's widow, Kay Truex De Justo, in Fresno, California, where she lives with her present husband, Michael De Justo. She is in her late fifties, a trim-looking, well-preserved brunette with a precise, no-nonsense style of speech that may reflect her educational experience—she was in graduate school in English at both USC and Brandeis University, although circumstances prevented her from obtaining a doctorate at either institution.

Kay told me that at the time she met Max in the summer of 1973, she was working as a teacher in Fresno, the city where she was born, but sometimes came down to Los Angeles to visit one of her brothers. On these visits she occasionally dated a colleague of Max's, but it didn't work out, and that colleague—whether out of kindness or in order to speed the end of the relationship—set up a blind date for her and Max. Although Max was eleven years older than she—he was thirty-seven and she was twenty-six—the two hit it off right away. Within a few weeks they went on a backpacking trip together in the eastern Sierras, and within four months they had married.

Kay loved Max's outgoing, lighthearted approach to life and his active lifestyle. Besides backpacking, he introduced her to skiing, which took them to Heavenly Valley at Lake Tahoe, among other locations. Max also continued to run; soon after they married they bought a house in the Hollywood Hills, so Max was able to run in Griffith Park, a slightly healthier environment than the Hollywood Freeway.

Children were quick in coming: Gene, their eldest son (no doubt named for Max's older brother) was conceived on their honeymoon and born in August of 1974, and their second son, John, was born two years later. A few months before John was born they moved to a new home in Manhattan Beach.

One evening in early 1979, when Kay was preparing dinner, she looked through the kitchen window and saw Max walking in from the garage as usual, but he was dragging one of his feet as he walked. Kay didn't think much of it; she knew that Max had injured that foot sometime during his running career. But it kept on happening, and after a few weeks it became his regular style of walking. Then she noticed something else: When she kissed Max good-bye in the morning, she saw that he had failed to shave a small part of his face, up by one ear. Max himself noticed problems. He had trouble raising his arms to wash his hair in the shower, and he also had voice problems. On one occasion when he was on his lunch break, he saw an old friend across the street and tried to shout his name, but no sound came out of his throat. These symptoms alarmed him, but he said nothing to Kay about them at the time, and he minimized the significance of the symptoms that she herself had noticed.

A few months later, one of Max's hands began to shake—a steady tremor that showed itself most when his arm was at rest. One of

Max's legal colleagues noticed the tremor and recommended that Max go see a doctor. He did so, and as a result Max was admitted to Encino Hospital. "They ran all manner of tests on him," said Kay, "and on Thursday of that week they told us that it was Parkinson's disease."

At the time he was diagnosed, Max didn't know much about Parkinson's disease, but Kay was well aware that it was a progressive and potentially fatal disorder of movement. "There had been a man in our church who had had it," she said. "You could just see him diminish. I knew it was very serious."

The central biological process that causes the symptoms of Parkinson's disease is thought to be the death of a set of brain cells that produce a neurotransmitter, or signaling molecule, called dopamine. These cells are located in a small region of the brainstem called the substantia nigra or "black substance"—so called because these cells are heavily pigmented. The cells have extensions called axons that run from the substantia nigra to the striatum, a structure higher up in the brain that helps to generate body movements. (More accurately, there are two striatums, one in the left and one in the right hemisphere of the brain.) The tips of the axons—the synapses—release dopamine into the striatum, and this release is vital for normal function.

In Parkinson's disease the cells of the substantia nigra gradually die over a period of years or decades, and their axons die too, so the striatum is gradually starved of its supply of dopamine. Well over half of the dopamine cells have to die before the disease shows itself, however: Thus, a person who experiences symptoms for the first time has actually harbored the underlying disease process for years without knowing it.

Of course, people who develop Parkinson's disease want to know why *they* got the disease rather than someone else. In most cases, the answer isn't known. Doctors commonly describe the disorder as "idiopathic," meaning that it seems to develop of its own accord without any obvious external cause. Still, there are some clues. In particular, chronic exposure to pesticides, herbicides, and other environmental pollutants raises the likelihood of developing the disease. What's more, some people have a genetic makeup that makes it hard for the body to break these pollutants down, and this makes such people more likely to develop Parkinson's disease if they are chronically exposed to pollutants.

Truex might have been exposed to agricultural pollutants during his childhood, when he lived in a farming area, drank well water, and ate fish caught by his father in a polluted lake. Also, as mentioned earlier, Truex was exposed to a variety of toxic agents in his adult life as a result of training and racing in highly polluted air. Still, it will never be known for sure whether these exposures were a factor in his developing Parkinson's disease, or whether his case was truly "idiopathic."

The mainstay of treatment for Parkinson's disease is the drug L-dopa. Once ingested, this drug enters the brain, where it is transformed into dopamine and thus makes up for the brain's own deficient supply. It is usually taken in a proprietary form called Sinemet, in which the L-dopa is combined with another drug that protects the body from some of L-dopa's potentially harmful side-effects. Truex did take Sinemet, and it helped him, but it did not prevent the progression of his disease. In particular, his speech began to be affected. He spoke too rapidly: When he was speaking in court, the court reporter would ask him to slow down, but somehow he couldn't. Then his voice weakened, so that it was hard to hear him in any kind of noisy environment, and he also began to slur his words. Max's brother Don told me that he took Max to task for "mumbling"—he didn't realize that it was a symptom of his disease. "I said, 'Max, you earn your living talking, you've got to talk so that you can be heard.'"

In spite of these problems, family life went on in a reasonably normal fashion. About six months after Max was diagnosed, their daughter, Mindy, was born. Max did as much as he could to help with the children. And he even kept on running. He now knew that his disease was likely to progress, but he focused on the hope of remaining an effective father and provider for as long as the children needed him.

Unfortunately, within two years or so his voice deteriorated to the point that he could no longer function effectively in court. Truex wanted to continue working by concentrating on office work, but the county thought otherwise: They retired him on a disability pension. The pension was adequate to maintain the family's standard of living, but the sudden termination of his career was a brutal experience for an active man like Truex. He worked for a few months in the office of a colleague, then quit working entirely.

Gradually, the disease started to close in on him. He began to have problems driving, especially when he needed to make rapid turns of the steering wheel. He had always looked forward to being an active role model to his children, taking them backpacking and skiing, teaching them athletics, and so on, but all these things became harder and harder. His doctor frequently increased the dosage of his drugs or added new ones, but they never quite kept up with the advancing disease. Max's walking became stiff and slow, and getting up from a chair required a great effort. He had to give up running—a terrible blow.

The Truexes thought that living in Los Angeles might be too demanding for Max, and they decided to move somewhere with a slower pace of life, where Max might be able to continue with activities such as driving. At the urging of an old Air Force teammate who lived in Gunnison, Colorado, they moved to that city. Both Max and Kay liked Gunnison. ("It was like being on vacation," says Kay.) But the move did nothing to slow the advance of Max's disease. He began to experience episodes of "freezing," when he simply got stuck in the middle of what he was doing and couldn't move at all. His walking became unsteady to the point that he was in constant danger of falling. The tremor in his arms worsened. And after he got into two car accidents in the space of two days, he had to give up driving for good.

Max remained under the care of his doctor in California, who juggled Max's drugs as best he could. Max was taking about thirty tablets a day, some of them intended to treat the disease, and some to counteract those drugs' side-effects. As often happens with people with Parkinson's disease, it became unclear which of his drugs were helping him and which were harming him, and so his doctor decided to take him off all his drugs for a couple of weeks so that he could be "releveled." The idea was to reintroduce the drugs one by one, while observing what effects they had. During the time that Max went without the drugs he stayed with Don in Santa Barbara, so Don got to see the full extent of Max's disease in its untreated state. "He came unglued," Don told me. "He couldn't do anything. He couldn't even swallow."

The releveling may have led to some temporary improvement, but Max soon began to go downhill again. By the mid-1980s he was having trouble with dressing, cutting up his food, and any other task that required delicate control of movement. His symptoms seemed to change from day to day. "If I was paralyzed from the waist down," Kay remembers him saying, "I would know what I could do and what I couldn't do, but with this I never know whether it's going to affect my ability to walk, my ability to speak, whether I'm going to shake or not shake, I never know which part of my body is going to go."

Max struggled on. As his sons got to the same age as he was when he started running, so did they. Max tried to coach them, but often Kay had to help out. Max's friends from his college and Air Force days rallied round. Kay recounts how one friend insisted on taking Max and his family to Vail for a skiing vacation: "He said, 'I don't care how many times you fall, Max, I'll pick you up.' And he did."

Toward the end of the 1980s, when Max had been battling Parkinson's disease for nearly a decade, things began to get rapidly worse. As so often happens with L-dopa treatment, the drug had begun to lose some of its effectiveness, and even with the addition of other drugs, he was in a serious plight. Now Kay had not only to cut up his food but to lift it to his mouth, too. And his swallowing was impaired almost as severely as when he had been off the drugs entirely: He would choke on his food as often as he swallowed it. Max became quite depressed and anxious about his future.

One person who followed Max's decline with great concern was his old track coach, Jim Slosson. When Slosson met Max in 1988, he was devastated to see how this brilliant athlete had been reduced to a shuffling invalid. A little later, Slosson was having dinner with one of his own old track buddies, a half-miler by the name of Paul Iacono, and he told Iacono about Max's illness. As it happened, Paul had a son, Robert, who was a thirty-six-year-old neurosurgeon with a special interest in Parkinson's disease, so Paul arranged for Max to go see him.

I visited Dr. Iacono in 2000 at his office in Redlands, California. He was a handsome and still young-looking man with black hair and a full black mustache. He certainly looked to be of Italian descent, as his name suggested. What was most memorable about him, though, was not his appearance but his manner of speech. He was unstoppable, and often very colorful in his choice of words. He seemed perpetually on the brink of saying more than would be wise. An interview that I expected to last for less than an hour went on for three hours, during which time I barely had the opportunity to get in a few questions. Various assistants and medical students drifted in or drifted out during our meeting, as if to get samples of their boss's oratory.

Back in 1988, Iacono was based at the University of Arizona Health Sciences Center and the Veterans Administration Hospital, both located in Tucson. He had gone to college and medical school at USC, and he had done further training at Duke University, where he had developed a special interest in the underlying mechanism of Parkinson's disease. "I was mostly grinding on new theory," he told me. "I wrote a couple of hundred papers when I was at the University of Arizona on neurobiological theory because I was trying to pave a new trail into the unmarked stuff. Because the marked stuff, the dogma, is just crap when it comes to Parkinson's disease."

If Iacono did write that number of papers, most of them must have remained unpublished, or else they were published in minor periodicals that are overlooked by the indexing services. But those that are available do attest to his iconoclastic approach to neurology during that period: They suggested novel causes and novel treatments for a number of disorders. In a paper published in 1990, for example, he suggested that a cluster of brain cells named the locus coeruleus played an important role in the development of Parkinson's disease. The neurotransmitter used by those cells is not dopamine but a related compound named norepinephrine (also called noradrenaline—it's a close chemical relative of adrenaline). If true, Iacono's hypothesis would suggest treatment options quite different from the traditional dopamine-related drugs used for the disease, such as L-dopa.

Thus, Max Truex was coming to visit a doctor with a very different mind-set from the conservative neurologists who provided his regular care. Furthermore, Iacono was not a neurologist at all, but a neurosurgeon. By that token, he might be expected to think in terms of dramatic onetime solutions rather than the painstaking manipulation of drug dosages over months or years.

Truex's visit was also different in the way he was treated. Whether because he had been referred by Iacono's father, because both men had been at USC, or because Truex, like Iacono's father, had been an outstanding runner, Iacono treated him much more as a family friend than as a patient. He picked Truex up at the airport, for example, and had him stay at his own home during his time in Tucson.

Although Iacono may have been a radical thinker, he says that it was Truex, not himself, who pushed for radical treatment. "I said, 'You look well medicated, I don't know what I can do for you,' but he said, 'No, Bob, I need something.'" Nothing was decided on during that visit, but Truex continued to call Iacono at his home on a more or less weekly basis for about a year. During this period the Truexes moved to Boston, so that Kay could enter graduate school at Brandeis University—she was hoping to restart her academic career and thus bring some more money into the family.

Truex himself seemed to be researching his treatment options during this period, because on one occasion he said to Iacono, "What about this adrenal graft?" This was a procedure popularized by doctors in Mexico, who took fragments from the patient's adrenal gland and transplanted them into the brain, near the striatum. Adrenal gland cells produce a certain amount of dopamine, as well as other chemicals, so the thought was that they might make up for the lack of dopamine in the striatum of people with Parkinson's disease. It turned out that the method didn't work—the transplanted cells may have survived for a while and maybe even provided some benefit during that time, but within a few months they died, and the patients were as bad off, or worse off, than they had been beforehand. Nevertheless, there was a short period of enthusiasm for the procedure right after the Mexicans published their findings in 1987, and Truex heard about it and proposed it to Iacono as a possible remedy for himself.

Iacono knew that the adrenal transplants were not working as well as advertised, so he talked Truex out of that particular option. But Truex still wanted something done. He kept calling and visiting Iacono. After a few months of this pressure, Iacono felt that he had to do something. So he decided that Truex should undergo a procedure that was even more radical than an adrenal transplant—a transplant of cells from the brains of human fetuses.

Fetal tissue transplantation as a possible treatment for Parkinson's disease had been pioneered by researchers in Sweden, led by Anders Björklund. In the late 1970s, the researchers had created an animal model of Parkinson's disease: They treated rats with drugs that destroyed their dopamine cells, which left the animals with a movement disorder that was somewhat analogous to Parkinson's disease. Then they took tissue from the substantia nigra of fetal rats—tissue that contained immature dopamine cells—and transplanted it right into the striatum of the treated animals. After a few weeks the animals' ability to move improved greatly; sometimes they seemed close to normal in their behavior. The transplanted dopamine cells had survived and matured in their new home, and they were supplying at least some of the dopamine that the rats were missing.

In late 1987 the Swedish group performed the same kind of transplants on two human patients with Parkinson's disease. The substantia nigra cells that they transplanted came from human fetal tissue obtained during abortions. Because the aborted fetuses were extremely immature—just seven weeks old and an inch or so in length—and because they were broken apart during the abortion procedure, it took a great deal of skill for the researchers to identify and dissect out the substantia nigra while leaving other unwanted tissues behind. But they accomplished this task successfully. They then drilled small holes in their patients' skulls, passed long hollow needles down through the cerebral cortex into the underlying striatum, and then pumped the fetal cells through the needles into the striatum. The patients recovered satisfactorily from the surgery, and the researchers observed them over the following weeks and months to see whether the transplants had any effect.

In fact, the patients did seem to improve. Although Björklund's group didn't publicize their research immediately, word spread through the neuroscience community. So Iacono heard about the Swedish experiments, and he suggested to Truex that a fetal transplant might benefit him, too.

Truex agreed; in fact, he leaped at the idea. Kay may have been a bit more cautious, but she went along with Max's wishes. "I was in favor because it was what he wanted," she said. And, though an observant Catholic, she had no moral qualms about the use of fetal tissue. "I didn't have a religious problem with it, because I knew that the tissue was going to be thrown away—it wasn't as if people were aborting children for this purpose."

The next question was where and how Truex was to get the transplant. Iacono first contacted the Swedish group, hoping that they would accept Truex as a patient. They turned him down, however, because they preferred Swedish patients who they could monitor for long periods after the transplant. Then Iacono found out that a group at Yale University was gearing up to perform similar transplants and was looking for volunteers. But, as it turned out, the Yale group also wanted people who lived locally. A group in England also turned Truex down. Iacono was stymied.

Then, in November 1988 the first American fetal transplant was performed by a team led by neuroscientist Curt Freed of the University of Colorado. Freed had researched the technique for years, first in rats (like the Swedes) and then in monkeys. This research was funded by federal grants, but when he began the human work, Freed had to turn to private funds because the Reagan administration, concerned about the abortion issue, had banned the use of federal grants to support transplantations involving human fetal tissue.

For the first transplant, Freed selected a volunteer by the name of Don Nelson, a fifty-one-year-old Denver man who had been suffering from Parkinson's disease for nineteen years. As with Max Truex, Nelson was deteriorating fast and he was desperate to try some new therapy. Freed obtained fetal tissue from an abortion clinic in the Denver area, dissected out the substantia nigra, and (with the collaboration of a neurosurgeon) injected the fetal cells into the striatum on one side of Nelson's brain.

If the Swedes were publicity-shy almost to the point of secretiveness, Freed was the very opposite: He held a news conference to announce the transplant just two days after Nelson's operation, long before he could know whether Nelson would experience any benefit from the procedure. Part of the reason for his haste may have been that the Yale group was about to do *their* first transplant—it actually took place just a few weeks later. In the world of medical research, priority is a significant issue.

Iacono did not contact Freed to see if he would accept Truex as a volunteer. Although I don't know the exact reason, the fact is that Freed and Iacono didn't get along. "I have no respect for Curt Freed," Iacono told me. "The results he's got have been so poor that no one should be continuing that work." And he described some of Freed's more recent experiments, in which he did mock surgery on some patients to establish a placebo control group, as "asinine and unethical." When I talked with Freed in 2000—we were collaborating on a book about Parkinson's disease—he was equally blunt about Iacono. He described Iacono as "one of the most, shall we say, 'provocative' neurosurgeons who has not been censured by the academy of neurosurgery but whom everyone has said *should* be censured." Of course, this was after he learned about what happened to Max Truex; it's possible that he was better disposed toward Iacono back in 1988.

In any event, Iacono said that he made great efforts to find a place where Truex could get a fetal transplant. "I tried all my friends all over the world—Sweden, Britain, Japan. I tried and tried. And eventually, after two or three years of following Max, I realized I had to do it myself."

Do the transplant himself? Iacono was a neurosurgeon, certainly, so he had a general expertise in brain surgery. He also had a particular interest in Parkinson's disease. But the handful of fetal transplants that had been done up to that time were performed by large research teams at major medical centers. The teams included basic neuroscientists, immunologists, and neurologists as well as neurosurgeons, supported by large amounts of money from public or private sources. The researchers had years of practice doing the transplants in laboratory animals before they ventured to touch a human being. They had studied every variable that might affect the success of the procedure, such as the right age of the fetus that would supply the tissue, the proper technology for dissecting, handling, storing, and administering the cells, and the right kind of drugs to give the patients to prevent the transplanted tissue from being rejected. The researchers performed brain scans, using advanced technology, to measure dopamine function in the brains of their volunteers before and after the transplants. And they carried out rigorous neurological testing of their volunteers for months before and after the transplants, so that any benefit or harm of the transplants could be assessed. All of this was overseen by university committees composed of doctors, administrators, and ethicists, whose role was to ensure that the volunteers were not exposed to undue risk, and that they were fully aware of those risks that could not be avoided. How was Iacono to replicate all of this, virtually by himself, and without any funds to support the project?

Iacono *was* based at a major medical center—the University of Arizona Health Sciences Center in Tucson—but he would not have been allowed to do the transplant there, for two reasons. First, the operation was outside his recognized area of expertise. ("I just don't do stuff like that," was how he put it.) Second, there was a general ban on medical research using human fetal tissue at the University of Arizona—a state school in a very conservative state.

Iacono thought that the best way to overcome these difficulties was to do the operation overseas. He first thought of Japan, which he visited from time to time. But his Japanese colleagues were reluctant to get involved. Then, while in Japan, he met a doctor who worked in a cancer hospital in Zhengzhou, the capital of Henan province in China. The doctor suggested Iacono perform the transplant there: Tissue from aborted fetuses was readily available, he said, and regulatory control was lax. "It's no muss, no fuss in China," as Iacono put it.

Traveling halfway around the world for a surgical operation is not unheard of. Plenty of people come to the United States from faraway places to have a surgical procedure that is not available in their home countries. And in 1989 a California neurologist took a patient with a Parkinson-like condition to Sweden for a fetal-cell transplant. (The Swedes were willing to bend their rule about not accepting foreigners in that case.) Still, what Iacono was proposing to do was very different: Rather than take Truex to an established center of excellence where the local doctors were experienced in the transplant procedure, he was planning to take him to what, in many people's minds, could be considered the "back of beyond," and more specifically to a hospital whose staff had absolutely no experience in this kind of surgery. Iacono was going to have to do almost everything himself, so if he was to succeed, he needed to be fully prepared.

Iacono did in fact prepare himself as best he could. "I was learning about immunosuppression, I studied up on the embryology, I read all the papers, and I developed my own technique—I solved millions of problems," he says. Still, he did not perform fetal-cell transplantations in animals as the other researchers had done, nor did he go and witness human fetal-cell transplant surgeries at one of the centers that were already doing them.

One problem stood out as the most challenging. The transplantation procedure involved stereotaxic surgery—that is, the use of a calibrated metal frame attached rigidly to the patient's skull. By mounting the injection needle on the frame at a specified location and angle, it could be driven into the brain a predetermined distance and the surgeon would know that the tip was in the desired target, the striatum. Iacono was familiar with the techniques of stereotaxic surgery, but such surgery couldn't be carried out at the Zhengzhou hospital—they simply didn't have the facilities.

Iacono thought up a fairly devious scheme to get around this difficulty. In April of 1989 he operated on Truex at the Veterans Administration hospital in Tucson. The operation was a "thalamotomy"—the destruction of part of a brain region called the thalamus. This is a procedure that is sometimes done to alleviate the tremor of Parkinson's disease, and in fact Truex's tremor was lessened, according to Kay. But the thalamotomy wasn't the main reason for taking Truex into the operating room that day. Rather, it was the "cover" (as Iacono himself put it) for a second procedure that he carried out "on the QT" immediately after the first.

In the second procedure, Iacono inserted three catheters, or plastic tubes, into Truex's brain, using the stereotaxic equipment that was available at the V.A. hospital. The tips of two of the tubes were guided into the left and right striatum. Iacono placed the tip of the third catheter in a ventricle—one of the large fluid-filled cavities inside the brain. It was the left lateral ventricle, which is close to the striatum on that side. This catheter was of a different design than the other two: Its back end was connected to a small rubber bladder, or reservoir, that Iacono implanted under Truex's scalp. Then he sewed up the scalp incision with all three catheters still in place. The idea was that later, in China, he would be able to push the fetal tissue down the tubes and he'd know that it would end up in the striatum or in the ventricle, even without stereotaxic control. The catheters would serve as pretargeted delivery chutes. What is more, Iacono thought that the inevitable tissue damage caused by the presence of the tubes would actually be beneficial to the transplant: Some research suggested that damaged brain tissue releases chemicals called growth factors that encourage cell survival.

Truex recovered uneventfully from these procedures, and a couple of weeks later the two men set out for China. They flew first from Tucson to Los Angeles. This was convenient for Kay, because the family had temporarily moved back from Boston to Manhattan Beach, where they were overseeing some work on their old property there, which they planned to sell. Thus Kay and Gene had the opportunity to come up to the airport and visit with Max and Bob during their layover. Max didn't look like a typical globe-trotting tourist: Besides his obvious Parkinsonian symptoms, his head was swathed in bandages to protect the locations where Iacono had drilled through his skull. "He was in pretty bad shape," said Kay.

At the airport, Iacono and Kay had a frank conversation. This is how Iacono recounted it to me: "I told Kay, 'You know, I may not be able to bring him back.' And she said, 'Bob, you've got to try desperate—' And I said, 'I may not be able to bring him back—even in a box.' She said, 'Bob, please try!' So this wasn't tiddlywinks. And I'm no Texas chainsaw murderer; I'm a very conservative neurosurgeon by the way." When I asked Kay about this, she at first denied any memory of such a conversation, but later she said, "I'm beginning to remember this 'box' thing. He probably did say something like that. He is very colorful."

The journey to Zhengzhou was a nightmare. Starting in Los Angeles, Truex and Iacono first flew to San Francisco, where they took a China Airlines flight bound for Shanghai. But fuel supplies ran low, and the pilot had to make an emergency landing in Japan. After a long delay they finally made it to Shanghai.

From Shanghai, they took a train for Zhengzhou. The five-hundred-mile rail trip took twenty-two hours, mostly occupied by repeated holdups as the steam-powered passenger train was forced to yield the track to higher-priority freight or military trains. And there was no food. Train travel in China was difficult at the best of times, but these times were far from the best: On May 4, about 100,000 students and workers had marched through Beijing to protest government policies, and this unprecedented event had greatly alarmed the government, so the entire country was in a state of tension.

Truex had been in a bad way at the start of the trip, but by the time they arrived at Zhengzhou he was virtually immobile: He could not walk even a few steps. Iacono had to carry him off the train; then he was put on a tricycle and wheeled to a waiting car.

Neither man had ever been to mainland China before, and Iacono was unprepared for the primitive conditions that existed at the Zhengzhou cancer hospital. There was no heat or hot water, for example, and the equipment was rudimentary. The microscopes, which were crucial for the dissection of the fetal tissue, didn't even have built-in light sources. Like children's microscopes in the West, they simply had little mirrors that you aimed at a window.

Another surprise had to do with money. According to Kay, the doctors at the Zhengzhou hospital, or the hospital administrators, demanded a substantial fee—she thought it was in the range of $20,000 to $25,000—to let the operation go ahead. Don said that Max told him the fee was close to the annual operating budget for the hospital. "I think they knew what they had," Kay commented, meaning that they had Max over a barrel. "Bob was a little taken aback."

One expectation was fulfilled, however: According to Iacono, fetal tissue was readily available. "I'd say, 'I need some things to dissect, guys, because I haven't had any practice in my country, bring me some stuff.' And it would be, 'You need fetal? OK, no problem.' And a couple of hours later they'd bring me something, and I'd say, 'Where did you find that?' and it was, 'Oh, in the dustbin.'"

In a medical paper describing the case, Iacono says that the samples used for the actual transplantations were obtained in accordance with NIH guidelines, which would include obtaining the mother's consent to the use of the tissue for transplantation. Max told Don that a female gynecologist rode through nearby villages on a bicycle, telling people what she was looking for. If so, she would presumably have had the opportunity to explain the planned use of the tissue to the women who had the abortions, and to get their agreement.

Some other research groups who were doing fetal-cell transplants, such as Curt Freed's, made an effort to tissue-match the fetal tissue to the recipient, at least at the relatively crude level of the ABO system (the system of antigens commonly known as "blood groups," but actually present in all tissues). Iacono did not tell me whether he did this, but it is unlikely that the Zhengzhou hospital had either the facilities or the expertise to carry out such an analysis. If they didn't, the compatibility of the fetal tissue with Truex's own tissue would be pretty much a hit-or-miss affair. Iacono simply assumed that the tissue would be a mismatch, and he started Truex on immunosuppressant drugs—specifically, steroids and a drug called cyclosporin. The hope was that these drugs would prevent Truex's body from rejecting the transplanted tissue, and he would need to continue to take them for the rest of his life.

Having taken these preliminary steps, Iacono started the transplant procedure, which consisted of three separate operations. On the first day he took tissue from a fetus that he judged to be sixteen weeks old. It's now known that dopamine cells from fetuses this old survive poorly after transplantation—by this age they have largely or entirely lost their ability to survive in a new host. But at the time, that may not have been so clear. At any rate, having obtained the fetal tissue, Iacono opened Truex's scalp and pushed fragments of the tissue down the implanted catheter that led to the striatum on the right side. Then he removed the catheter and closed the scalp incision. It was a very brief operation.

A day or so later Iacono repeated the procedure, again using tissue from a sixteen-week-old fetus. This time he implanted the tissue in Truex's left striatum. He used a somewhat different technique: He loaded the tissue into a small spring-shaped metal coil and pushed the entire coil down the catheter into the brain tissue; then he removed the catheter, leaving the tissue and the metal coil in place.

For the final procedure, Iacono used tissue from a much younger fetus—it was sixteen millimeters long, which would make it only five or six weeks old. Strictly speaking, it was still an embryo and not yet a fetus. This age was younger than what is now considered the optimal age for transplantation, and the tiny size of the fetal brain at that age made accurate dissection a major challenge. Anyway, Iacono dissected out the tissue he wanted, broke it up into tiny fragments, injected it into the reservoir attached to the third tube, and squeezed the reservoir to force the cells down the tube and into the lateral ventricle. After doing so, he left the third tube and reservoir in place, because he planned to use it to take samples of Truex's cerebrospinal fluid at later dates.

Injecting the cells into the ventricle, rather than into the substance of the brain, was a novel and risky step. Iacono did this with the hope that the cells, once in the ventricle, would secrete chemical "growth factors" that would in turn promote the survival and growth of the cells he had injected into the striatum. He says that there were animal experiments to support this hypothesis. But the ventricles of the brain are linked together to form a single tortuous waterway through which the cerebrospinal fluid circulates. So putting cells in the lateral ventricle was rather like throwing alien water plants into a particular pond in the Everglades: One might expect them to spread widely through the system, with unpredictable and possibly harmful consequences.

Iacono told me something even more remarkable about the transplants he performed on Truex. He didn't take the fetal tissue from the brain region that everyone else was using—the substantia nigra—even though that was the location of the dopamine cells. "That just shows you my contempt for the dopamine hypothesis," he said. Instead, he took tissue from a strip of the brain near the midline, running from near the front of the brain all the way back to the medulla, where the brain narrows down to join with the spinal cord. This zone contains a diverse assortment of cells that use neurotransmitters other than dopamine. As mentioned earlier, Iacono thought that some of these other cells and chemicals played a more central role in Parkinson's disease than the dopamine system. Iacono sticks to the same ideas today, even though dopamine remains the chief culprit in most other people's opinions.

Truex came through the three procedures without any problems, and within a few days Iacono decided that it was time to go home. But that was easier said than done. On May 20 the Chinese government, in response to the increasing unrest and continued demonstrations, declared martial law. All regular forms of transportation were halted. "I had to give up all my cash and my passport for several days," said Iacono, "and they finally dragged four people kicking and screaming off a Russian turboprop [to make room for us]. I was at the point of tears by then. The tires of that plane were worn down to the Dacron. But we finally got home, and three days later I heard that that very same plane had crashed and killed everyone onboard." Kay added another detail related to her by Max: On June 4, the two men passed through Beijing's Tiananmen Square in a pedal-powered taxi on their way to the international airport. Just hours later, the massacre began that cost the lives of hundreds or thousands (accounts differ) of students and workers.

Truex finally rejoined his family in Manhattan Beach. He was very tired, but hopeful that the cost and exhaustion of the trip would pay off for him. And, according to Kay, he had good things to say about all the people who cared for him in China. In that connection, Jim Slosson added a fairly implausible detail that he supposedly learned from either Truex or Iacono. During his hospital stay in China, Slosson said, Truex spent his time running up and down the corridors after the Chinese nurses. "The good-looking ones," Slosson added with a wink.

Before long the family moved back to Boston. According to Iacono, Truex's condition improved greatly over the eighteen months that followed the surgery. Already at six weeks his facial expressions were livelier, and soon thereafter his speech and his gait improved. By fifteen months he was able to resume normal daily activities, Iacono says, and he even began running again, in order to coach one of his sons. Iacono described the overall improvement as "so dramatic." He did however, offer one caveat, saying that no one had checked to see if the immunosuppressant drugs that were used to prevent rejection of the grafted cells might themselves have a beneficial effect on the symptoms of Parkinson's disease. When I asked him to expand on this far-from-the-mainstream notion, Iacono clammed up. "I'm not going to tell you that stuff," he said. "I have to write my own book about that."

Iacono wasn't alone in his belief that Truex did well after the transplant: Other people who knew Truex before and after the operation were of the same opinion. "There was nothing dramatic or immediate," said Kay. "I remember, I think it was about a year after surgery, Bob called to see how he was doing, which he did periodically, and I said, 'You know what, I think Max is doing very, very well, because it just occurred to me that for about the last three weeks to a month I haven't been doing anything for him, he's been doing everything for himself.'" Don Truex agreed. "His facial expressions were better, his speech was better," he said. "There's no question in my mind that he was substantially improved." Jim Slosson also said that Max was better after the transplant.

Iacono was impressed enough with the results that he took other patients to China for similar transplants, but he performed the surgery in a more modern hospital in Shanghai rather than in

Zhengzhou. He described some of these cases, including Truex's, at several scientific meetings that he attended.

In the second year after the surgery, Truex visited a neurologist by the name of Raymon Durso, who is a specialist in Parkinson's disease at the Veterans Administration hospital in Boston. Durso also has an academic appointment at Boston University Medical School.

"I think I saw him a total of three times," Durso told me. "He definitely said he was improved. However, when I went over the history, his doctors had, for example, added Deprenyl, and so I was never willing to attribute his improvement to the surgery." The FDA approved Deprenyl for the treatment of Parkinson's disease in the same month that Truex received his transplants. In some patients it significantly alleviates the symptoms of the disease, and it can also cause euphoria, so that a person may feel and act more upbeat even if the physical symptoms are unchanged.

The last time Truex visited Durso was in March of 1991, a year and ten months after the trip to China. Durso says that Truex seemed quite well at that time, aside from some swelling of his ankles. Such swelling can be caused by mild kidney failure, and Durso confirmed this diagnosis by means of blood tests. Chronic mild kidney failure is a common finding in people who are treated with immunosuppressant drugs.

Although he seemed reasonably well to Durso, Truex evidently did not seem well to himself, because right around that same time he began telling Kay that he was dying. Kay didn't take him seriously at first, but he was insistent. When she asked him what made him say that, he replied, "I just know, I just don't feel the same, I just know I'm dying." Then, over a period of ten days or so, Max began giving Kay specific instructions concerning his death. "He told me about his life insurance policy," Kay says. "He had me go into Boston and sell off some stocks that he didn't want to cause problems afterwards, and he had me write down that he wanted to be cremated, and where he wanted his ashes." He also gave Kay power of attorney for medical matters, and told her that he did not want to be resuscitated if the question arose. Kay realized that

Max was very serious about his belief that his life was ending, even if he couldn't verbalize the reasons.

Two weeks after the visit to Durso, on the morning of Sunday, March 24, Kay woke to find that Max was already out of bed. She got up and went downstairs in her nightgown, and she found Max in the living room. This is her account of what followed: "I asked him if he wanted me to fix him some breakfast, but he said, 'No I've already had something. I'm kind of tired, I think I'm just going to lie down on the couch for a while.' I wasn't feeling well either, so I lay down on the other couch. The kids weren't up yet. I was dozing, and then I heard him moving around, and it was like he caught himself on the arm of the couch, so I got up and I said, 'Do you need help, are you OK?' He said the strangest thing—it was like something changed, something was going on in his head—and he said out of the clear blue, 'I think I can still pee in a can.' And then he just kind of slumped. I looked up, and my son John had come down the stairs and was right there. By then we were in the doorway between our living room and our kitchen, and I said to John, 'Grab a chair quick, Dad's going down.' So he came over and we got Max into a chair, and he kind of slumped over and it was like he was snoring: a puff and a puff and a puff and then a rush of air out of his lungs. And I said, 'John, stay with Dad, I'm going to get dressed, I'm going to take him to the emergency room, I don't know what's happening.' And I came back down as soon as I could and said, 'How is he?' John said, 'Fine, Mom, he's sleeping.' But he wasn't sleeping."

Kay called an ambulance, but it took twenty minutes or so to arrive. Kay knew that Max was dead. Gene and Mindy came downstairs, and the four of them talked about what had happened. When the paramedics finally arrived and began to insert a tracheal airway, Kay remembered what Max had told her just a few days previously, and said, "He doesn't want this." The paramedic said, "Do you have something legal?" and Kay explained about her power of attorney. Then the paramedic took out the tracheal tube and said, "He's been gone for some time, it would have been much too late anyway. I was doing it mainly for the children"—presumably, to assure them that everything possible had been done to save their father. The ambulance crew left, after telling Kay to call a funeral home and have them collect the body.

Even though she and the three children were in a state of extraordinary shock and grief, Kay remembered something else that Max had told her during his final days, which was that, when he died, she should contact Iacono, because he would want Max's brain saved for scientific study. And when Kay Truex called Iacono later that Sunday to let him know that Max had died, he was indeed very eager to have Truex's brain examined. He knew that it might be possible to detect the presence of the transplanted cells. To show that these cells had survived for nearly two years after the transplant surgery would be an important scientific finding, and it would provide a partial validation for the fetal-tissue treatment. There was no reason to suspect, at that point, that the transplants had anything to do with Truex's death—he might easily have died from some other unrelated condition such as a heart attack. So, after some calls between Iacono and Dr. Durso, Kay was asked to have Max's body taken to the New England Medical Center.

She did arrange for the funeral home to take him there, and Kay and the children followed the hearse in their own car. When they got to the hospital, the staff, who had been alerted to the situation, had Max's body taken to the morgue, and Kay returned home. In the face of her own and her children's grief, she had to alert other members of the family. She called Don, leaving it to him to break the news to his and Max's mother—a widow of ten years and now the mother of two sons who had died suddenly and unexpectedly.

I asked Kay whether Max's last words, "I think I can still pee in a can," meant anything to her. She said that it seemed to be just a random fragment recalled from his childhood. "I know that his mother had told me that she potty trained the boys by having them pee in a can," she said. "They liked the noise."

In the regular way, brain autopsies are leisurely affairs. Within a day or two of the person's death the brain is removed and placed in a bucket of formaldehyde, where it sits for several days or weeks until it has hardened sufficiently that it can be easily sliced and studied.

But this was not to be the regular way. Iacono wanted to use very sensitive chemical procedures to detect the presence of the transplanted cells. These methods identified certain enzymes present in those cells—enzymes that were responsible for synthesizing the particular neurotransmitters that those cells produced. For these procedures to work, the brain tissue had to be as fresh as possible.

Iacono says that he asked Durso to arrange for the brain to be removed that same day. Durso began making phone calls to locate someone who would be able to do the procedure. He tried several neuropathologists who he knew at the Veterans Administration hospital, but because it was a Sunday, and already late in the day, he wasn't able to track anyone down. Then he tried to page other neuropathologists around town, and finally, during the evening, he reached Rebecca Folkerth.

Folkerth was on the staff at the New England Medical Center, but she wasn't on duty at the hospital that day, or even on call. "But I was one of the crazy foolish people who leave their beeper on all the time," she told me. "I answered my page on the Sunday night and I said, 'OK, I'll come and do this autopsy.' It sounded like Durso was having trouble getting anyone to help him."

Folkerth reached the hospital around nine p.m. "Once I got there, I got a call from Dr. Iacono," she says. "He told me the whole history and said, 'Can I ask you to take some of the tissue fresh and freeze it?' I said, 'OK, fine.' It's not the usual thing—we usually put it in formalin and let it harden for a couple of weeks."

So Folkerth donned scrubs and a face shield, identified Truex's body, wheeled it out of the cooler, and began the procedure. First, she placed a block under Truex's head, raising it a few inches to make it more accessible. Then she took a scalpel and made a long curving incision in Truex's scalp, starting behind one ear, passing over the top of the head, and ending behind the other ear. This separated the scalp into front and back halves. She took hold of the front half and pulled it down over Truex's face, and then she pulled the back half backward and down over his neck, leaving most of his skull exposed. Then she took a power saw and began to cut off the entire top of Truex's skull. Even with the power saw it was hard physical work, and it took about thirty minutes. In usual circumstances, the job is often left to the *diener,* the technician who runs the morgue.

Having removed the skullcap, Folkerth cut the cranial nerves and the blood vessels that supply the brain, and then sliced across the top of the spinal cord so that Truex's brain was now entirely separated from the rest of his body. Being in its natural, unhardened state the brain was jelly-like and difficult to handle, but Folkerth placed it in a dish, took a long broad-bladed knife, and sliced the brain as best she could into a series of slabs, each about a half-inch thick.

Up to now, Folkerth hadn't noticed anything unusual about Truex's brain. "But as I was cutting it," she told me, "I made this observation, 'Gee, look at this strange stuff in the ventricles, in the third and fourth ventricles, and in the lateral ventricles also.' I thought, 'Isn't that odd?' and I took a bunch of pictures. And I thought, 'That looks like cartilage; isn't that weird!' Even to the naked eye it looked like cartilage, and there were hairs—you could see them, just eyeballing it—the gross pictures are extremely dramatic." By "gross" pictures, she meant the pictures she took with a regular camera, as opposed to pictures taken through a microscope. She didn't mean that they looked gross, though in fact they did.

Brains don't usually contain cartilage or hair, of course. Nor bone or skin, which she later discovered were also present. "You could see the hair shafts," she went on. "So I knew there was something very strange about this right away. Oh, this was the most strange thing I'd ever seen, and at this point it was the middle of the night. I was the only one there, looking at this case and thinking, 'What the hell is this?' It was creepy. So here I am taking these pictures and thinking this is some mistake; this is a tumor—a teratoma."

A teratoma is a tumor derived from embryonic stem cells that retain the capacity to form many or all of the body's various tissues. Most commonly, teratomas form in the ovary, but they can be found in other places and in either sex. Teratomas contain a chaotic mixture of tissues, which can include cartilage, skin, hair, bone, gut, retina, brain, glands, even teeth. It's as if the tumor is trying to form a fetus, but without any conception of how the various tissues are supposed to be arranged.

It looked like something similar had been happening inside Truex's head. Lumps of glistening cartilage lined the floor of one of the ventricles. Part of one of the lateral ventricles was completely filled with a waxy, skinlike tissue. The fourth ventricle, which is located in the brainstem near nerve centers concerned with breathing and other vital functions, was packed full of hair and other tissues, so much so that some of the surrounding brain structures were compressed and discolored.

A teratoma? Folkerth knew that that was exceedingly unlikely. Teratomas can on rare occasions occur in the brain, but if so it would usually be in the brain of a newborn infant or young child. A brain teratoma in a fifty-three-year-old man would be a diagnosis of desperation. Then Folkerth remembered the case history that Iacono had recounted to her—the Parkinson's disease and the transplants. And as she examined the brain slices further, she began to figure things out. In the left striatum she found the metal coil that Iacono had used to convey the second transplant into Truex's brain. And she saw the catheter that Iacono left in place after the third transplant. The tip of the catheter was still located in the left lateral ventricle. "I thought, 'It can't be a tumor; it's the tissue they infused in there.' There was no other explanation."

Folkerth froze some specimens of tissue for later microscopic examination, as Iacono had requested. She put other parts of the brain in formalin. Then she reassembled Truex's head as best she could: She replaced the skull cap, pushed the scalp back over it, and sewed the incision roughly together. Having returned his body to a more-or-less lifelike appearance, she wheeled it back into the cooler. The next day it was collected by the funeral home, and later it was cremated in accordance with Truex's wishes.

This whole experience left a big impression on Folkerth, and so over the next few months she devoted a lot of her free time to analyzing the tissue samples from Truex's brain. In the left and right striatum, where Iacono had deposited the tissue from the two sixteen-week-old fetuses, she found no surviving cells from the transplant, only scar tissue. This was consistent with findings from other research groups, who have reported that tissue from fetuses this old has a very low chance of surviving the transplantation procedure. Folkerth concluded that the reported improvement that Truex had experienced was not due to the presence of any transplanted nerve cells in his brain. Either just the damage caused by the injections had a beneficial effect, which didn't seem terribly likely, or some other factor, such as the new drug that Truex received, was the reason.

What about all the weird tissues in the ventricles? These presumably arose from the tissue that Iacono had dissected from the very young, five- to six-week-old fetus and had injected into the left lateral ventricle. Folkerth believes that Iacono mistakenly included some tissue that was not from the embryo's brain at all—tissue from just outside the brain that normally would have developed into the overlying bone, cartilage, skin, and hair. Those cells could have drifted through Truex's ventricular system, found some attachment point, multiplied, and followed their own normal developmental pathway, unaware that they were now in a highly inappropriate location.

I asked Folkerth whether she thought that the blockage of the ventricular system was the cause of Truex's death. "In my heart of hearts, yes," she said. "I think that was the cause, but it wasn't a complete autopsy so I can't rule out a heart attack, pulmonary embolus, or something like that. The story that his wife told me made it sound like he had respiratory failure. I think he had gradual changes in the brainstem [where breathing is controlled] that couldn't be compensated for any longer, because we saw a lot of chronic changes, microscopically."

Iacono had been the initial driving force behind the autopsy, and it would have been natural for him to participate in publishing the findings that emerged from it. In fact, at a scientific meeting three months after Truex's death, he announced that the results of the autopsy were "pending." But later, Iacono seemed to lose interest in having the results published. And that wasn't too surprising, perhaps, because the findings suggested not only that two of the transplants had failed to survive, but also that the third had survived only too well, and had quite likely caused Truex's death.

Still, Raymon Durso and Rebecca Folkerth felt that the findings should be published, because at that time there were only one or two autopsy studies of fetal transplant recipients, and the results in Truex's case seemed to offer an important warning to researchers in the field. So, after some delay, Folkerth and Durso decided to write the paper on their own without Iacono. For the clinical details of the case they would rely on what he had told them and what he had reported at that scientific meeting.

After more than a year's delay, they sent their manuscript to the *New England Journal of Medicine,* because that journal had already published several articles about fetal-cell transplantation for Parkinson's disease. But the manuscript was rejected. "That was funny," says Folkerth. "I thought this was something that was definitely worthy of being in that particular journal. There seemed to be kind of a pro-transplant point of view in the other articles they had published."

What Folkerth didn't know was that her manuscript was reviewed by Curt Freed, a major enthusiast for fetal-cell transplantation and an author of one of those "pro-transplant" articles in the *NEJM*. As he later told me, Freed recommended that the manuscript be rejected. The reason was a concern that, even though it only described what he considered a "therapeutic misadventure," it could bring the entire procedure into disrepute. (Nevertheless, Freed has had his own setbacks with the procedure. Three years after Truex's death one of Freed's patients suffered a brain hemorrhage during the transplant operation; he died a few weeks later.)

The rejection of the manuscript caused another delay, but in 1995 Folkerth and Durso sent the manuscript to another, less prestigious journal, *Neurology.* It was accepted, and it appeared in 1996, five years after the autopsy it described. Folkerth and Durso didn't name Iacono in the body of their article. "I didn't want to indict the guy, I didn't want to be too accusatory," Folkerth says. Still, they did thank both Iacono and Kay Truex in a footnote, so anyone in the field would have realized which case they were talking about.

Iacono didn't respond to the *Neurology* article, or if he did his response didn't get published. But the journal did publish a response from a research team that had begun to do fetal-cell transplants at the University of South Florida in 1993. Evidently this team, like Curt Freed, was worried that Folkerth's article would throw the field of fetal-cell transplantation into disrepute, and they expressed their feelings about what Iacono had done in unusually strong language. "This is a case of extremely poor tissue dissection," they wrote. "One wonders why this transplant was performed in China," they added, "outside of State and Federal regulations, Institutional Review Board oversight, and peer review scrutiny." "We should not be surprised," they concluded, "that poor science leads to poor outcomes."

Iacono remains convinced to this day that Truex was greatly helped by his transplants, and he doesn't believe that the tissue in his ventricular system caused his death. "There weren't any signs of increased intracerebral pressure," he told me. "He wasn't having urinary incontinence, he wasn't showing signs of dementia, he wasn't complaining of headaches. He was acting normally, and his wife said he came in and sat down and died. That just doesn't sound like [ventricular blockage]. His death was officially signed out as a heart attack." (Kay says that Max's death certificate lists only "Parkinson's disease" and does not mention any immediate cause for his death.)

A few months after Truex's death, a memorial service for him was held at USC; it was attended not only by family members but also by many of Truex's old teammates from his college and Air Force days. Jim Slosson was there too. As a more lasting memorial, his family and friends endowed a college scholarship for athletes from Warsaw High School. There is also a Max Truex Memorial interscholastic track meet that is held in Indiana every May.

Max's mother, Lucile, died exactly nine months after Max, on Christmas Eve of 1991. She had been in frail health, but the shock of her son's death accelerated her own, Kay believes. Kay stayed on in Boston for a year so that Gene could graduate from high school, and then she and the younger children moved back to Fresno, the city of her birth. In 1993 she attended her thirtieth high school reunion, and there she ran into Michael De Justo, a classmate she had been out of touch with for decades. Within a few months they married.

During all the years since Truex's death, neither Kay nor anyone else in the family learned what Rebecca Folkerth found in his brain—not even after her findings were published. Kay tells me that she did have a phone conversation with Folkerth some time after her husband's death, but all she learned from that was that he had not suffered a stroke. As to whatever else Folkerth said during the conversation, Kay said, "I could not for the life of me understand what she was saying to me." Thus it is possible that Folkerth did describe what she found, but did so in technical language that failed to communicate much to a layperson like Kay.

It wasn't until the summer of 2005, when I met with Kay in Fresno, that she learned about what had happened and saw Folkerth and Durso's published report. She was of course surprised to learn that none of the fetal brain cells had survived, and shocked to see the photographs of the nodules, hair, and other fetal tissues that were growing in the ventricular system of Truex's brain.

Kay did take issue with one thing in the report. Folkerth and Durso, citing Kay as their source, had written in the summary of their report that Max had died after a "several-hours interval of progressive lethargy and breathing difficulties"—a description that would be very compatible with an impairment of brainstem function. "That is completely incorrect," said Kay. She reiterated that Max had not complained of tiredness until a few minutes before his death, and had not shown any breathing difficulties until the very last moments of his life. "I think they went back after the fact," she said, meaning that Folkerth and Durso misremembered or misrepresented what Kay had told them in a manner that fit in better with their pathological findings. To be fair to Folkerth and Durso, the main text of the report does not state that Truex had breathing difficulties for hours prior to his death, but only tiredness.

I had thought that Kay might react to what she learned with considerable hostility toward Iacono, but she didn't—not in the couple of hours I was with her, at least. On the contrary, she reemphasized her belief that Iacono had acted out of good intentions and that Max himself had urged Iacono to go ahead with the procedure. "If this [report] is true, it's very sad in a way," Kay commented, "because it means that what Max set out to do to help himself may have actually gone completely the other direction."

Iacono stopped doing fetal transplants in 1989, after he had operated on a total of twenty-five to thirty patients, all of them in China. "When you start adding up the negative aspects of fetal grafts," he told me, "including the risks of immunosuppression as well as infection from the fetus and contamination from these other things, the risks of fetal grafts are pretty high." In a paper published in 1994, Iacono argued that fetal transplantation was a less successful treatment for Parkinson's disease than another neurosurgical procedure called pallidotomy, which involves destruction of part of a brain region called the pallidum. At the time I visited Iacono he was specializing in pallidotomy operations: He did them, as he put it, in "industrial numbers."

Some other centers, such as Curt Freed's, continue to perform the transplants, with mixed results: About one-third of the patients have been greatly helped, some have seen little change in their condition, and a few have developed disabling side-effects of the procedure, such as involuntary flailing movements. In Freed's hands the transplanted cells do survive, and no patients have been afflicted by the teratoma-like growths that Max Truex experienced.

In the waiting room of Iacono's Redlands office I noticed a life-size portrait of a surgeon operating, with a man standing next to him who is guiding his scalpel. Oddly, that man is wearing neither gown nor mask nor gloves. It took me a moment to figure out the reason: That man is Jesus. Iacono has become quite religious since the Truex days, and he no longer approves of abortion or of using aborted fetal tissue for science. "I went from 'I don't care what I'm doing here with a fetal graft' to becoming a right-to-lifer," he said. "I'd see these little guys, and after a while you realize you can tell how they're going to grow up and what their personality's going to be like; you can almost name them."

Since my meeting with him in 2000, things have not gone well for Iacono. In October 2001 the Loma Linda University Medical Center, where Iacono was doing his surgery, revoked his privileges, meaning that he could no longer operate there. According to the California

Medical Board and newspaper accounts, the hospital's action was provoked by a laundry list of misbehaviors, starting in 1992 with an episode of "inappropriate language and inappropriate touching." In 1994 Iacono allegedly used some drugs that were not approved by the FDA. This was followed in 1998 by "yelling and abusive behavior toward staff," which earned him an official reprimand and six months of anger-management therapy. In May 1999 Iacono is said to have become angry with a scrub technician in the operating room, and to have grabbed her hand, causing an injury. In the spring of 2000, according to the allegations, he told a nurse, within earshot of a deceased patient's family, that she had "killed" the patient. And at some unspecified time, Iacono is accused of having allowed a medically unqualified nurse practitioner to drill holes through patients' skulls.

Following the loss of his surgical privileges at Loma Linda, Iacono applied for privileges at another hospital, Desert Regional Medical Center in Palm Springs. But, according to the California Medical Board, Iacono falsely answered "no" to a question about whether he had ever had his surgical privileges suspended or revoked. Because of this and the other alleged actions by Iacono, the Medical Board brought a formal accusation against him in 2004, and in September 2005 Iacono was ordered to surrender his medical license, meaning that—unless he can gain reinstatement—his medical career in California is over. His license in Arizona is still valid, but Iacono currently lives in Mississippi, and as far as I know he does not practice medicine any longer.

Why did Truex agree to participate in a project that he must have realized was hazardous in the extreme, and which quite likely killed him? Why did he agree to be operated on by someone who had absolutely no previous experience in this kind of work, in an absurdly remote location, and without any kind of regulatory control? In part, of course, it was simply his desperate desire for relief from his incurable and progressive illness. But also, he placed a great deal of trust in Iacono. He was a family friend, after all. And Iacono, whatever failings he may have, is an extraordinarily vivid and persuasive talker. At our meeting in 2000, after lecturing me for several hours Iacono left the room and a medical student who had been sitting in on our meeting turned to me and said, "You haven't seen him at his finest. He gets very dynamic—a very charismatic fellow!" Then Iacono popped back in and said to me, hopefully in jest, "My Mafia friends can track you down and cut your tongue out if this doesn't work out for us."

CHAPTER 2

METEOROLOGY:
All Quiet on the Western Front

"Earlier on today apparently a woman rang the BBC and said she heard that there's a hurricane on the way," said Britain's best-known weatherman, Michael Fish, on an October evening in 1987. "Well, if you're watching, don't worry—there isn't." Then the hurricane struck. Eighteen people died, and losses mounted into the billions. Michael Fish should have been hanged.

That, at least, is a synopsis of events as they have engraved themselves in the memory of Britons who survived the Great Storm of October 15 and 16, 1987. Fish himself sees it all a bit differently. He was talking about the weather in America. The videotape was doctored. The French let him down. His forecast wasn't wrong. He wasn't even on duty that evening. And there was no hurricane.

Untangling what actually transpired before and during the storm presents quite a challenge. For one thing, meteorology is an arcane science. Take the term "baroclinic instability," a key concept in the analysis of the 1987 storm. A baroclinic instability is the condition when isobaric-isopycnal solenoids are nonzero. Got it? I think not.

Also, as befits an arcane science, its practitioners stick together like glue. The debacle of October 16—if it was a debacle—was followed by an orgy of collective amnesia in which any hint of scientific failure—if there was a failure—was masked, denied, or erased from the record. It wasn't till more than twenty years later that a trio of (non-British) meteorologists broke ranks and laid out how the forecast *should* have been done.

———

One thing is clear: there *was* a hurricane—in America. Its name was Hurricane Floyd, but it shouldn't be confused with the monster of that name which devastated the Carolinas twelve years later. (That one led to the permanent retirement of "Floyd" from the christening book of tropical storms.) By comparison, the Floyd of October 1987 was a timid thing—in fact it carried official hurricane status for only a few hours on the night of Monday, October 12, as it traversed the Florida Keys. Then it weakened to a tropical storm, crossed the Bahamas without causing unusually severe damage, and petered out.

Floyd didn't make it as far as Europe, but its influence did. By October 14, energy propagating downstream from the remnants of the hurricane had helped establish a deep east–west trough of low pressure in the northeastern Atlantic, off the coast of Spain. To the north-west of the trough, frigid air descended from the arctic; to the south, warm, moist air was being carried northeastward over Spain. The resulting steep north–south gradient in temperature was the ultimate energy source for the storm, making it an "extra-tropical cyclone." (Depending on their location, storms of this type may be referred to more specifically as "mid-latitude cyclones" or "European wind storms.") A true hurricane, in contrast, is born in tropical waters and feeds off the *vertical* temperature gradient between the warm ocean surface and the cold air aloft.

I discussed the events of 1987 with Thomas Jung, a German research meteorologist based at the European Center for Medium-Range Weather Forecasts (ECMWF) in Reading, England. Jung and his colleagues have re-created, within the safe confines of a modern supercomputer, the ferocious conditions of the Great Storm.

"When you have these temperature gradients, the wind increases with height," Jung said. "This is the so-called geostrophic wind, which produces a very large shear, and a baroclinic instability develops, and then you have development of strong storms." Explaining this a little: A geostrophic wind is a wind that blows in a direction parallel with the isobars—the lines of equal pressure that are the main element of weather charts. You might think the wind would turn left, cross lots of isobars, and fill up the low-pressure trough.

Indeed, the pressure gradient is trying to make it do exactly that, but it doesn't succeed, because another, equally strong force is trying to make the wind turn right. That other force is the Coriolis force that results from the Earth's rotation. The two forces balance out and so the wind simply charges straight ahead *along* the isobars. The wind's speed is proportional to the pressure gradient: That is, it gets faster as the isobars become more closely spaced.

Near the Earth's surface, however, the wind is slowed by friction and turbulence. This causes the Coriolis force (which is proportional to speed) to become weaker, so the wind in the lower parts of the atmosphere *does* turn somewhat to the left, crosses isobars, and runs partway down the pressure gradient. Thus there is a shearing action between the slower, leftward-turning wind at lower levels and the faster, straight-ahead wind at higher altitude. The shearing action generates an unstable situation called a baroclinic instability, which involves a complex interaction between pressure, density, and temperature at different locations and altitudes. I confess that I don't understand the details, but the outcome is a vortex—a counterclockwise spinning action that winds up low-pressure troughs into the high-energy systems called extra-tropical cyclones. These are often visible as comma-shaped cloud systems in satellite images of the Earth's middle latitudes.

Extra-tropical cyclones are good things, because they very efficiently convey heat and moisture from the tropics toward the poles, thus making living conditions more tolerable for everyone. Still, the fact that they originate in instabilities is a problem, especially for weather forecasters. An instability is a knife-edge-like situation. "If there is a small perturbation, it grows and grows and grows," said Jung. "The problem is that if this happens with the atmosphere itself, it is likely also to happen with any error in your forecast."

People think of the task of weather forecasting as getting from the current conditions to the conditions tomorrow or the day after. But that actually omits a crucial first step, which is getting from the observational data to the current conditions. Forecasters call this step producing an "analysis"—a mapping of pressure, temperature, humidity, wind strength and so forth—for the region of interest at the time the observations are made. It may be presented as a two-dimensional chart such as one sees on television, but the vertical dimension of the atmosphere is equally important. Why it's called an analysis, I don't know—it's really more of a synthesis.

The analysis is always inaccurate to some degree. That's because the observations themselves are not perfectly accurate, and because they don't provide complete coverage. The latter problem is particularly severe for a country like Britain, most of whose weather comes off the empty expanses of the Atlantic Ocean. Up until the 1970s, Britain and other countries maintained a fleet of about ten weather ships that were stationed at various positions in the North Atlantic. Between bouts of seasickness, the crew would release balloons that ascended to the stratosphere, radioing back the atmosphere's vital signs as they climbed.

The ships gradually fell victim to fiscal belt-tightening, however, as well as to the belief that satellite-based meteorology was the wave of the future. That was a mistake, according to Jung, because satellites have a hard time measuring conditions at different vertical levels in the atmosphere. At any rate, by the time of the 1987 storm, weather ships were history. Surface observations were restricted to those provided by a few automated buoys and by whatever commercial ships happened to be in the area of interest. These buoys and ships didn't release weather balloons. What's more, some of the commercial ships were so notorious for providing inaccurate data that they were put on an official blacklist of untrustworthy sources. Some of the missing data was supplied by commercial aircraft, but most of the transatlantic jetliners traveled to the north of the Bay of Biscay, the birthplace of many European storms, including the one of October 1987.

As if these problems weren't bad enough, another factor may have made the data-acquisition process even worse than usual during the days leading up to the 1987 storm. Michael Fish told me that the meteorologists in France were on strike, and as a result they failed to send weather data from their country to the Meteorological Office (or Met Office) in Bracknell near London, which was then the nerve center of British weather forecasting. Bill

Giles, who was Fish's boss at the BBC Weather Center, confirmed this account to me. There is no mention of any such strike in the news reports that I read, however, nor in the official report that was published in the aftermath of the storm, so I sent an inquiry about it to Météo France. I got a reply from Jacques Siméon, who was head of the forecasting division there until 1987. He told me that a one-day strike of the entire French Civil Service was called for October 15, but that few people in Météo France heeded the call, and as a result there was no effect on forecasting or the transmission of data.

Having developed a representation of the current conditions—an analysis—forecasters have to predict how things will change in the future. By 1987 the Met Office had largely consigned this task to computers. In fact, the Bracknell center already possessed a "supercomputer"—a CDC Cyber 205—capable of executing 200 million calculations per second. Still, this is only about one-millionth the speed of the fastest present-day supercomputers. Even at the time, it was outmoded in comparison to the Cray supercomputers being operated by the French meteorological service and by the ECMWF in Reading.

The Bracknell forecasters used two different models to transform the analysis into a prediction of future weather. To simplify the computing process, both models represented the atmosphere as a three-dimensional grid of boxes, each of which was assigned a single value of temperature, pressure, and so on. In one model, the boxes were about 60 kilometers on a side. This so-called fine-mesh model was restricted to the United Kingdom and surrounding areas. The other, coarser model, used boxes of about 120 kilometers on a side. This UK global model as it was called, covered the entire Earth. Both models divided the vertical dimension of the atmosphere into about twenty levels. (By comparison, the most detailed present-day models use boxes as small as 4 kilometers on a side, and they divide the atmosphere into about seventy levels.)

Applying the equations of fluid dynamics to the data in each box and its neighbors, the supercomputer changed the data in each box to represent the passage of some small increment in time. This was done repeatedly until the computer had arrived at the time of interest—say, twelve or twenty-four hours after the starting time. The result was a data set that human forecasters could use to draw charts and predict future conditions.

Several days before the storm struck, it was already clear that bad weather was heading toward Britain and France. Starting on Sunday, October 11, television and radio forecasts warned that wet and stormy conditions were likely toward the end of the week. This was not good news to Britons, who had experienced exceptional amounts of rain over the previous few weeks, with flooding in low-lying areas throughout southern England. By Wednesday, the actual weather over England still gave no hint of trouble to come, but the warnings handed out by the Met Office had grown more urgent. The lunchtime BBC television forecast on that day included mention of a low-pressure system that "is going to deepen like mad and head up and give us an angry spell of weather, wet and windy" on Thursday and Friday.

Even while the weatherman was speaking, however, things were starting to go wrong back at the weather operations center in Bracknell. When the fine-mesh model was applied to the data for noon on Wednesday, the result was—no storm at all! A broad area of low pressure would cross the country and might drop some rain, but winds would be light. The global model did still show a storm, with a track crossing southern England, but the low-pressure center was not deep enough to signify that exceptionally strong winds were on the way.

Meanwhile, out in the real world, conditions were rapidly deteriorating. By Wednesday night, the trough of low pressure in the Bay of Biscay had developed into a classic extra-tropical cyclone, and it was racing toward northwestern France and the western approaches to the English Channel. Its central depression was deepening rapidly, even explosively, and the polar air that had been heading south was whipping around the depression in a counterclockwise direction and heading back toward the northeast. Ahead of it was a great mass of warm, subtropical air that had been bitten off and now formed an isolated, onrushing lobe of heat and moisture. Thus there were two fronts heading northeastward: a warm front marking the leading edge of the subtropical air and, a couple of hundred miles behind it, a cold front marking the leading edge of the pursuing arctic air. (An animation of the advancing weather system is available online—see the list of sources for this chapter.)

At about two a.m. on Thursday, the Met Office's Cyber 205 began chewing on the data that described the weather situation at midnight, using the fine-mesh model. After an hour or so it spat out the result: Yes, the depression would be deep enough to cause stormy conditions after all, but the center would track up the English Channel, perhaps grazing the very tip of southeast England (the county of Kent) on its way to the North Sea and the Low Countries. This seemed like good news for Britain, because it meant that the right-hand side of the depression, where the strong winds were likely to occur, would affect the English Channel and northern France, sparing the British mainland.

This tendency for the strongest winds to occur to the right of the center of an extra-tropical cyclone is a phenomenon that is also seen with hurricanes. It is caused by the fact that, on the right side, the speed of the entire advancing system adds to the speed of the cyclonic circulation, whereas on the left side it subtracts from that speed. This effect is not very pronounced with hurricanes, whose eyes typically progress quite slowly—perhaps 10 mph. But extra-tropical cyclones move much faster—the center of the cyclone that caused the Great October Storm raced across Britain at more than 50 mph. This caused a very large difference between wind speeds on the left and right sides of the storm track: To the left, gentle breezes wafted from the northeast; to the right, hurricane-force gales blew from the southwest.

Having finished its work on the fine-mesh model, the Met Office's computer turned its attention to the global model. Because it was now about three a.m., some more recent data could be fed into the model that was not available for the running of the fine-mesh model. Whether on account of the new data, or for some other reason, the output of the global model was quite different from that of the fine-mesh model. It predicted that the center of the depression would veer left, make landfall in southwest England about midnight, and cross the country well north of London during the small hours of Friday morning. What was more, the model predicted that the pressure at the center of the depression would be 965 millibars. This was 48 millibars below mean atmospheric pressure and, in combination with the predicted steep pressure gradient to the south of the advancing center, it would be enough to generate very strong winds in the region of England to the southeast of the storm track.

Thus, when the Bracknell forecasting team for Thursday, October 15 came on duty, they were faced with a quandary: Their two computer models predicted utterly different conditions for the following night. Either southeast England would experience a major windstorm, or it wouldn't.

This was a situation where the task of weather forecasting was suddenly thrown back into the laps of human beings, and specifically into the lap of the senior forecaster on duty that day. (The position is now called chief forecaster.) I haven't been able to find out that person's name. Ewen McCallum, a senior forecaster who was present in the operations room on that day (but not on duty), wouldn't tell me; others I spoke with professed not to know, and the reports issued after the storm didn't mention him by name. Evidently there has been some desire to protect the identity of the person who (in terms of his official status at least) was most responsible for the erroneous forecast. As McCallum commented, "it was a 'there but for the grace of God go I' kind of thing."

McCallum did give me one insight. The then head of the Central Forecasting Office (and thus the boss of the senior forecasters) was a man by the name of Martin Morris. During the previous week, McCallum told me, Morris had been skeptical of the idea that a storm was on the way, in spite of the models that indicated that it was. "He stressed to me that he was worried that we were perhaps overdoing things [in our forecasts]," McCallum said. On Thursday, Morris got involved in discussions with the forecasting team, something that was not necessarily part of his day-to-day job. "I have no idea how much pressure he put on the senior forecaster, but I certainly think there were discussions taking place," said Mc-

Callum. This seemed to be a broad hint that Morris urged the senior forecaster to opt for a milder forecast than he might otherwise have done, but I haven't been able to confirm this story from other sources.

In any event, the forecast that went out to the media on Thursday morning was a compromise between the results of the two models. It described the depression as likely to move up the Channel and across southeast England, with very high winds in the Channel and windy, but not exceptionally violent, conditions in the southeastern counties. It also spoke of the likelihood of considerable rain.

It was on the basis of this guidance that Michael Fish issued his infamous "no hurricane" forecast. Although many sources, including the Met Office's own Web site, describe Fish's forecast as having been delivered on Thursday evening, it was actually given at 1:25 p.m. as part of the BBC's lunchtime television news. This lunchtime broadcast was not very widely watched. Thus, when Fish told me that he was "not even on duty at the time," he presumably meant that he didn't give the more popular evening forecast—it was his boss, Bill Giles, who handled that one. Still, it was the crucial two sentences from Fish's broadcast, endlessly replayed during the days and weeks after the storm, that became emblematic for the Met Office's failure to properly predict the storm.

I was curious to know whether Fish regarded his job as requiring him to help develop weather forecasts, or whether it was more a matter of presenting forecasts generated at Bracknell. He was emphatic that his role included actual forecasting, and that all the BBC weather presenters were trained Met Office scientists. Knowing where I was calling from, he couldn't resist a dig at his U.S. counterparts. "The situation in America is pretty appalling," he said. "The weather should not be entertainment; it's a life and death kind of thing." Bill Giles made the same point: the BBC presenters were independent forecasters, he said. He, too, bemoaned the trend, especially evident in the United States, toward weather forecasts as entertainment. "Television wants nubile young ladies," he said, "and meteorology likes fat old gray men with experience."

Fish's and Giles's disparagement of the U.S. television forecasters may not be entirely fair. A good number of American weather presenters, probably including some of the nubile ones, have undergraduate or advanced degrees in meteorology or related disciplines, according to the Web site of the American Meteorological Society. Fish himself learned his trade while serving tea to the weather forecasters at Gatwick Airport, starting in 1962. He doesn't have any college degree, although in a 2004 interview posted on the Web site of London's City University he was quoted as saying that he earned a degree in physics at that institution in 1968.* Still, the BBC Weather Center has always had the reputation of being a scientific organization, not just a group of presenters parroting forecasts prepared at Met Office headquarters.

Given that Fish mentioned the entertainment issue, it may be appropriate to say a word about his physical appearance at the time of his famous broadcast. A staple of TV forecasting since 1974, Fish was a balding, mustachioed forty-three-year-old, who liked to wear thick, dark-rimmed eyeglasses and wool sweaters under plaid jackets. He sported a collection of ties with fish motifs and (if the *Sunday Herald* is to be believed) his underwear was also personalized, this time with weather-chart symbols. Fish had a love-it-or-loathe-it kind of style; fittingly, he was once voted Best Dressed Man *and* Worst Dressed Man on television in the same year.

Of course, I wanted to know the story behind the woman who called in to ask whether there was a hurricane on the way. This is a question that Fish has been asked many times before. In 2004, when he was interviewed for an article on the BBC Weather Center's Web site, he answered it as follows: "Nobody called in. . . . My remarks referred to Florida and were a link to a news story about devastation in the Caribbean that had just been broadcast. The phone call was a member of staff reassuring his mother just before she set off there on holiday."

---

*When I inquired about this, the university's alumni relations assistant, Christina DeMercado, told me that Fish did not earn a degree there, and she deleted the relevant passage from the Web page. The version that included Fish's degree claim remained viewable on another page, www.city.ac.uk/marcoms_media/print/default_print_1_6194_6194.html.

Fish gave me pretty much the same account, and he went on to bemoan how the often-shown video clip had been edited to make him seem as if he was talking about the weather in England. "If you had the complete clip there it would be painfully obvious it was nothing to do with the situation [in England]," he said. "The rest of the broadcast went on to say, 'Batten down the hatches, there's some extremely stormy weather on the way.' Which to me is a very good forecast."

This account is at least partially incorrect, according to a study of the television and radio forecasts before the storm that was published in 1988 as part of the report of the official investigation. Fish didn't say "Batten down the hatches . . ." on that broadcast at all. He uttered that remark, or something like it, in the course of a different forecast that he gave thirty minutes later. This was a forecast for the European satellite television Superchannel. His exact words were "It's a case of batten down the hatches, I think, for some parts of Europe; some very, very stormy weather on the way indeed." In other words, Fish was saying that exceptional winds would occur over continental Europe—something that all the models were agreed on—and not that they would affect England. Later in that broadcast he specified France and the Low Countries as the areas at risk.

On the BBC broadcast Fish did say that it would get very windy, but he gave less emphasis to the wind than to the prospect for rain, in line with the existing concern about flooding. The charts that accompanied the television forecasts that day indicated sustained wind speeds of up to 50 mph, but only for the English Channel and North Sea, not for land areas.

What about the woman who called in? Was she really the mother of one of his colleagues who was planning a trip to Florida? Not at all, according to the *Daily Mail*. In the aftermath of the storm, that enterprising tabloid posted a monetary reward for the name of the woman involved. The answer soon came in: It was a Mrs. Anita Hart from Pinner, one of London's northwestern suburbs.

I tracked down Mrs. Hart and spoke with her by telephone in 2006. "Oh no, no!" she cried in mock despair when I told her why I was calling. "This has been haunting us for the last twenty years."

She told me that she saw the *Mail*'s reward offer but didn't respond to it because she valued her privacy. "But we were shopped [turned in for financial gain] by one of our son's friends. To get the reward he called the newspaper and identified us."

Mrs. Hart's story started a few days before the storm. She and her husband were planning a trip to Wales with their caravan ("or RV, as you call it"). Their son Gaon was then studying meteorology at Manchester University. "We were in the habit, if we wanted to go away for the weekend—we would phone him and ask what the weather was going to be like. On this particular occasion he said, 'Don't laugh, but I think there's going to be a hurricane.' He had tapped into the French computers, because our computers were not up to it."

This was on Monday. By Wednesday there were still no storm clouds on the horizon, so Anita called the BBC Weather Center and asked whether there was indeed going to be a hurricane as she'd been warned. A man whom she took to be Michael Fish (but who probably wasn't) replied, "No, we don't get hurricanes in England."

When I told Mrs. Hart of Fish's explanation for his remarks on the broadcast—namely that he was talking about Florida, and that the video clip had been edited to make it seem as if he was referring to England—she laughed again. "That's absolute nonsense. That's not true at all. Obviously his story changed."

Although Fish's version of events can hardly be true, it has established itself as authentic in many quarters, including the BBC Weather Center's Web site, the Wikipedia article on the storm, and so on. Thus many sources portray Fish as the innocent victim of a hatchet job by the media, when in reality he himself handed them the hatchet.

Fish's broadcast did not endear him to his superiors. In talking with me, Bill Giles described Fish's comments as "stupid" and "silly." Ewen McCallum seconded the "stupid" and added "dumb" for good measure. The official report published in the following year described Fish's comments as "particularly unfortunate." It wasn't that Fish was single-handedly responsible for the bungled forecast, but that by issuing such an unqualified denial of the danger, he made any kind of defense of the Met Office's performance impossible. "The Fish thing was very important because that was the damning piece of evidence," said McCallum. "That was the 'No further questions, Your Honor.'"

Soon after Fish delivered his broadcast, the Met Office's Cyber 205 embarked on yet another round of number-crunching, using the weather data for noon. This time, the results of the fine-mesh and global models were in reasonable agreement, and they also agreed with the "compromise" forecast that had been issued earlier. This may have caused the Bracknell forecasters who had settled on the earlier compromise to congratulate one another on their good judgment, but if so their satisfaction was short-lived.

At any rate, they issued a forecast that was similar to the earlier one, and consequently Bill Giles's presentation on the Thursday evening news was similar to the one given by Michael Fish at lunchtime—minus the hurricane story, of course. Again, the emphasis was on the prospect for heavy rain, with some mention of strong winds in southeastern coastal areas. Thus most Britons went to bed without an inkling of the disaster that was about to strike. If they took any precautions, it was to station their wellies (rubber boots) by their front doors, not to batten down any hatches.

Around the time of Giles's evening presentation, the center of the oncoming depression lay in the western approaches to the English Channel. It still seemed as if the Met Office's forecast—a track up the channel and across the southeasternmost portion of England—was a good possibility. Between seven p.m. and midnight, however, the depression began to veer onto a more northerly path, and the depth of the depression intensified. By midnight the depression was nearing the coast of Devon in southwest England, and the sea-level atmospheric pressure at its center had decreased to about 953 millibars. This corresponded to the pressure that (on a normal day) would be encountered at an altitude of about 1,500 feet. It was at least 12 millibars below any estimate offered by computer models during the run-up to the storm. (By comparison, the pressure in the eye of Hurricane Floyd—the famous Floyd—dropped to 921 millibars.)

With the extreme deepening of the depression, the winds picked up to very high speeds. The highest winds were far above the land surface and thus were never actually measured. From the spacing of the isobars to the south of the storm track, however, it can be calculated that "geostrophic" winds at altitude would have blown at about 300 knots (345 mph). The actual high-altitude winds would have been blowing at somewhat less than geostrophic speeds, because they would have been slowed by centrifugal forces as they followed the curved path of the isobars. Still, they must have been blowing at speeds far above the threshold for hurricane status (65 knots, or 75 mph) or even the threshold for the most violent, category 5 hurricanes (135 knots, or 155 mph). Of course, as Michael Fish stated, this did not mean that the storm was in fact a hurricane.

At the Earth's surface, winds were much slower, but they were still exceptionally strong. In the English Channel to the south of the depression, and along the adjoining English coast, winds reached 50 knots around midnight, making it a "severe gale" or "storm" on the Beaufort scale of wind strengths. A gust of 95 knots (109 mph) was recorded on the French coast at about the same time.

During the small hours of Friday morning (October 16) the center of the depression began to track across southern England, well to the north of its forecasted path. The depression lay over Bristol around three a.m., Birmingham around four a.m., Nottingham around five a.m., and Hull (on England's North Sea coast) around six a.m. These cities experienced only gentle winds, or even complete calm, as the depression passed over them, but it was a different story to the south. Steady winds of 60 knots or more were widely reported, and gusts reached much higher speeds. The strongest gust recorded on the English mainland was 106 knots (121 mph) at Gorleston on the coast of Norfolk. Even in London, a gust of 94 knots (108 mph) was recorded at the British Telecom tower at 2:40 a.m., shortly before London's electricity supply failed and the anemometer stopped recording. This was the highest wind speed ever recorded in the capital. The highest wind speed measured anywhere during the storm was a 117-knot (134 mph) gust recorded on the coast of Normandy shortly after midnight.

Besides the wind and rain, southeast England experienced remarkable changes in temperature as the warm front and the pursuing cold front passed overhead. Shortly before midnight at the Met Office's station in South Farnborough, southwest of London, the temperature rose by 9°C (16°F) in twenty minutes, and fell by the same amount a few hours later. Rapid pressure changes also occurred, particularly as the tail end of the depression swept across England: Rises in pressure by as much as 12 millibars in one hour were recorded toward the end of the night. Both the temperature and the pressure changes were unprecedented in terms of their rapidity.

The human impact of the storm was felt earliest in the Channel. A catamaran carrying six people was put into difficulties off the coast of Dorset: The Weymouth rescue lifeboat set out in mountainous seas and, with support from a Royal Navy destroyer, was able to rescue the crew. A coaster, or small cargo ship, was damaged off the Isle of Wight and required rescue by two lifeboats. Farther to the east, it was even worse. A bulk carrier capsized and sank outside Dover harbor, taking the lives of two crew members. A cross-Channel freight ferry, the *Hengist,* ran aground near Folkestone, and the crew had to be brought to shore by an antiquated rope-and-harness contraption known as a breeches buoy. (The ship was saved and now plies the calmer waters of the Aegean under a different name.) Two passenger ferries, with 400 passengers aboard, were unable to enter port and had to ride out the storm at sea. A lifeboat set out from Sheerness, on the north coast of Kent, to save a fishing boat that was running aground; it rescued the crew but then was itself driven aground by a violent gust and had to remain there for the rest of the night.

Onshore, sixteen people died. The most frequent cause of death was treefalls, which killed nine people in England. One of these, a homeless man, was killed directly by a tree that fell in a London park. Four people—including two firemen returning from an emergency call—were killed by trees that fell on their vehicles, and three others died when their vehicles collided with fallen trees or while swerving to avoid one. Another person, a woman in Chatham, Kent, died when a tree crashed through the roof of her home.

Other deaths included three people who were killed by falling chimney stacks, a fisherman killed by a flying beach hut, and a motorcyclist who was blown into a barrier by a violent gust. Besides the British victims, four people died in France.

The main physical damage was to trees: An estimated fifteen million of them were destroyed over a span of about six hours. Two factors conspired to maximize the damage. First, most deciduous trees were still in leaf, thus giving the wind an easy purchase. Second, the ground was sodden from weeks of rain, making the trees much less securely rooted than usual. Not even those trees whose roots held fast were safe, however: Many of them succumbed to trunk breakage, a mode of failure that greatly reduced their salvage value.

In falling, the trees wreaked havoc. Many fell on their owners' homes. Here is an excerpt from an account written by a meteorologist, H. D. Lawes:

> We gave up trying to sleep and I made a cup of tea. . . .
> Up to a point one feels that your own home is the safest
> place to be but I had never seen any storm like this and
> I was beginning to feel uneasy. . . . Another particularly
> strong gust started up, there was a huge rumbling crash
> and a rushing noise. The house shook, the ceiling cracked
> and outside masonry was falling. We realised that the
> cedar tree had fallen and rushed upstairs to where baby
> Sam had been asleep. There was a gaping hole in the wall
> of his bedroom and the ceiling light was blowing in the
> wind. Luckily Sam was still there in his cot with only a
> few pieces of plaster and brick on him. He was crying
> lustily, which was reassuring, and Sue plucked him from
> the cot. We rushed downstairs and out of the front door,
> pausing only to rescue the cat, my pipe and tobacco. Outside we had to clamber through a morass of branches and
> foliage to where a neighbour was holding a torch [flashlight] and we gratefully accepted their hospitality.

Trees fell on roads and rail tracks in unimaginable numbers, making them impassable. Among the millions of Britons who found themselves trapped in their homes or unable to get to work were Anita Hart and her husband. At least they felt grateful that their son's hurricane warning had caused them to cancel their holiday. "Our caravan would have blown over," Anita said. Indeed, many caravan parks near the coast were scenes of complete devastation.

Trees also fell on power lines, causing outages. In fact, by 3:30 a.m. nearly the entire southeast of England, including most of London, had been plunged into darkness that lasted until dawn. With nowhere to send their electricity, power stations had to shut down abruptly. Many communications links were broken, either by trees falling on lines or by the collapse of broadcasting towers.

Although the immediate harm of the treefalls lay in the damage they caused, it was the sheer loss of trees that most affected Britons in the longer run. Southeastern England had not experienced such a violent windstorm in nearly three centuries; thus many trees had had time to grow to maturity and beyond without ever being severely tested by the elements. Majestic, ancient trees were a beloved and seemingly permanent feature of the English landscape, but that changed in the Great October Storm. Among the places hard hit were the Royal Botanical Gardens at Kew in West London, Britain's premier arboretum and botanical garden. Five hundred trees—about one-third of the garden's stock—were destroyed, and an equal number were severely damaged. Among the trees lost were many rare specimens, such as a cherry-bark elm (*Ulmus villosa*)—a species that is being harvested out of existence in its native Kashmir. Even greater damage occurred at the Royal Botanical Gardens' second site at Wakehurst Place in Sussex, where more than half the trees were downed. Other arboretums, parks, private gardens, and woodlands suffered major losses. There was also severe damage to commercial forests: Losses equaled about two years of lumber production for the entire United Kingdom.

Treefalls were the most prevalent type of damage done by the storm, but homes also suffered damage from chimney falls, stripping of roofs and sidings, and the like. One in six homeowners in southeast England filed insurance claims after the storm, and total insured losses (for both Britain and France) amounted to $4.2 billion.

For meteorologists in particular it was a memorable night. Some were at their homes and lived through experiences like the one described by H. D. Lawes earlier. One meteorologist lamented that he slept soundly through the weather event of the century—like a California geologist sleeping through the "Big One" on the San Andreas Fault. Other meteorologists were on duty that night. Those at Bracknell had to cope with a power outage that halted the Cyber 205 in its tracks around four a.m. Although any forecast for mainland Britain was now superfluous, the computer was also programmed to provide individualized, automated forecasts for the oil rigs in the North Sea. Thus, as the storm roared toward the rigs, the night staff had to hand-process, type, and fax the forecasts—an operation that took many hours. (One rig narrowly escaped disaster as a ship that had broken away from its moorings bore down on it: The ship was pulled away just in time.)

Most of the Bracknell day staff failed to show up, so the night staff had to continue on duty. One person who did show up was Ewen McCallum. He had prepared a summary of the week's forecasting for a conference that day. This task had been making him nervous all week because he was new on the job and also because there had been no weather of interest to talk about. When he arrived, his boss said, "Never mind the bloody conference, the shit has hit the fan and you're going to be on in an hour to explain it!"

In fact, it soon became clear that the Met Office would have a lot of explaining to do, because the storm provided a field day for the media, especially the tabloid press. WHY WEREN'T WE WARNED? screamed the headlines. The *Daily Express* spoke of the "complete, shameful devastation of the credibility of those smugly useless TV weather people." Even the more staid *Independent* spoke of the "dead silence from the frog-spawn watchers," and the *Guardian* described the Met Office staff as being "at the top of everyone's list of duffers." Many papers ran stories alleging that the French and Dutch meteorologists got the forecast right, unlike their incompetent British counterparts.

The quotes just mentioned are taken from an analysis of press responses published by Met Office staff. The analysis also mentioned that the Met Office received hundreds of letters from members of the public, the overwhelming majority of which expressed "support for the Office and appreciation of its services." This is a bit too self-serving to believe, but Britons do have a reputation for siding with underdogs. At least one highly critical letter did arrive. It was from the BBC, which accused the Met Office of damaging the corporation's reputation by failing to provide adequate warning of the storm.

Luckily for the Met Office, the Great Storm was quickly driven off front pages by an even less heralded disaster. The following Monday was Black Monday, when stock markets collapsed around the world. The London market lost £50 billion in total share value on that day alone, and by the end of the month more than a quarter of the London market's value had been wiped out. Many Britons whose homes had escaped the wrath of the storm now faced even worse damage to their pocketbooks.

Of course there had to be an investigation into what went wrong with the storm forecast, and who better to conduct it than the Met Office itself? In fact, the person who was given the main job of reviewing the forecasting process was the head of Central Forecasting, Martin Morris—the very man who, according to Ewen McCallum, may have pressured the duty forecasters to tone down their warnings. Not surprisingly, the report laid the blame on a factor that no one at Bracknell could be held accountable for—the shortage of data from the Atlantic. In fact, about the only person who was singled out for criticism was Michael Fish (for his "particularly unfortunate" no-hurricane comment), but even he wasn't mentioned by name.

By way of bolstering their assertions about the cause of the erroneous forecast, the meteorologists described how they repeated the run of the fine-mesh model that had been most in error (the one that ran after midnight on October 15 and that influenced Fish's "no hurricane" forecast at lunchtime). This time they added the extra data that had not been available at the time. According to the report, the model now came up with a good prediction of the storm's track and intensity. But the researchers admitted that they achieved success "by selecting values of the adjustable parameters." When researchers model events that happened in the past, almost any model can be made to predict the actual outcome if one twiddles the knobs for long enough: This, it seems, is what the Bracknell group did.

Perhaps aware that an internal investigation might not satisfy the Met Office's critics, the Secretary of State for Defence commissioned a review of the report's findings by a well-known mathematician, Peter Swinnerton-Dyer, and a university-based meteorologist, Robert Pearce. These two were somewhat freer in their comments than were the authors of the internal report. They criticized the duty forecasters at Bracknell for following the computer models too closely and for failing to recognize a situation in which the models were likely to underestimate the strength of the winds. Swinnerton-Dyer and Pearce also compared the British Meteorological Office unfavorably with its French and Dutch counterparts. The French meteorologists, they said, were better educated and trained and had faster computers, and as a result did a better (though still imperfect) job of forecasting the October storm. (Jacques Siméon of Météo France confirmed this to me.) Some of Swinnerton-Dyer's and Pearce's recommendations were implemented: Most notably, the Met Office was promised a new supercomputer, which was finally installed in 1991.

Thomas Jung, the ECMWF scientist who recently reanalyzed the 1987 storm, believes that weaknesses in the Met Office's computer models, and not problems with the data, were responsible for the erroneous forecasts. When Jung applied the ECMWF's current model to the then-available data, he got an excellent prediction of the track and intensity of the storm for several days beforehand (though for some reason the prediction broke down during the final few hours before the storm). Jung tested the sensitivity of the model's prediction to slight variations in the data, such as might be caused by erroneous or absent measurements at some locations. He did this by running the model fifty times, each time using slightly different starting conditions. The model robustly forecast the storm in spite of these perturbations. This kind of "ensemble forecasting" is now standard practice, because it gives the forecasters a measure of how confident they can be in the model's predictions. Thus it helps avoid situations like the one that occurred in 1987, when the models were in conflict but there was no objective way to assess which one was closest to the truth.

How did the Great October Storm measure up to other historic windstorms? The previous great storm of this type to strike southern England occurred on November 26 and 27, 1703. The effects were described by novelist Daniel Defoe (of *Robinson Crusoe* fame) in his 1704 book *The Storm, or a Collection of the Most Remarkable Casualties and Disasters which happen'd in the late Dreadful Tempest both by Sea and Land*. The storm followed very much the same track across southern England as did the 1987 storm and was probably similar in magnitude, but it did much greater damage on account of lower building standards in those times. Some details recorded by Defoe, such as the stripping of the lead roof of Westminster Abbey, suggest that winds may have been even stronger than in 1987. About 8,000 people died: These included the Bishop of Bath and Wells, who was "found with his brains dash'd out," and 1,500 sailors who drowned when a dozen ships of Queen Anne's navy were sunk.

Actually, windstorms nearly as powerful as the 1987 event cross the British Isles quite frequently, but they nearly always affect areas much farther to the north, such as the Hebrides Islands or the Cairngorm Mountains of Scotland. These attract very little attention because they cause next to no damage. "They just blow the sheep backward," as Bill Giles put it. Remarkably, another very powerful windstorm swept across England just twenty-seven months after the Great October Storm. The storm of January 25, 1990, was accurately forecast, but because it struck during the day and affected a wider area, it caused more casualties: Thirty-nine people were killed in Britain and about fifty in Europe. Another three million trees were destroyed in Britain, adding to the toll of the 1987 storm.

The most disastrous European windstorm on record occurred on January 16, 1362. Known as the Grote Mandrenke or "Great Drowning," it raised a wall of seawater that surged for miles across the coastal lands of the Netherlands and neighboring countries. At least 25,000 people drowned and the coastline was permanently altered. The loss of life would doubtless have been much greater, had not the Black Death already killed off about half the population.

The Great October Storm is now receding into history. A generation of Britons has grown up with no memory of the storm itself, nor of how southern England looked before it struck. Perhaps an occasional young person may wonder why Ashdown Forest barely merits being called a forest, or why a town with but one ancient oak would call itself Sevenoaks, but that is it. And gradually, new trees are replacing the old.

Michael Fish retired in 2004. In spite of his performance on the day of the Great Storm, he received numerous honors for his long service to the BBC. These included an honorary degree from City University, which may have made up for his lack of an earned degree at that institution. The Queen named him a Member of the Most Excellent Order of the British Empire. And after leaving government service he was free to capitalize financially on his fame—or notoriety. A few months later he was on the radio advertising central heating systems. "A woman rang to say there was a hurricane on the way," he told listeners. "Well, I couldn't care less, because I have this new wonderfully efficient Worcester central heating boiler, so whatever happens I will be as warm as toast and have loads of hot water." In 2006, perhaps yearning to be back in front of the cameras, Fish tried out for Independent Television's amateur song competition, *Celebrity X Factor*. He failed to win a spot after Simon Cowell poured faint praise on his rendition of "Singing in the Rain."

CHAPTER 3

※

VOLCANOLOGY:
The Crater of Doom

Volcanoes demand respect, but they don't always get it. In 1993, when geologist Stanley Williams led a party of scientists to their deaths in the crater of an active volcano, he triggered an eruption of controversy and blame.

The volcano in question was Galeras ("the galleons"), a 14,000-foot-high peak in the *cordilleras* of southwestern Colombia. On its eastern flank, just four miles from the summit but 5,000 feet below it, lies the city of San Juan de Pasto. Although it has a population of more than 300,000 and is the capital of the Colombian department of Nariño, Pasto is a fairly sleepy provincial town that is largely cut off from the bustling metropolitan centers to the north. If anything keeps the citizens of Pasto on their toes, it is the rumblings and occasional eruptions of their local volcano.

Galeras presents a real danger, in the sense that it has erupted frequently over recorded history. Significant eruptions occurred in or around 1580, 1616, 1797, 1830, 1865–1869, 1891, and 1936. With a record like this, one always has to keep the possibility of an eruption in mind, especially with a fair-size city so close by.

Galeras is also dangerous on account of the type of magma (molten volcanic rock) that it produces. Known as andesite, Galeras's magma is rich in silica and consequently is thick and pasty, especially after it is exuded onto the surface and has a chance to cool. Unlike the more liquid magmas found in Hawaiian volcanoes, which run smoothly down the volcanoes' outer slopes as incandescent lava flows, the magma at Galeras tends to pile up where it erupts, forming solid domes of lava that eventually seal off the vents through which the magma reached the surface. Thus, if pressure continues to build as more magma is forced upward from below, the result may be a sudden and difficult-to-predict explosion.

Frequent, explosive eruptions are dangerous, but there are also factors that tend to lessen the hazards posed by Galeras. For one thing, the historical eruptions have been fairly small—most of them have been confined to the volcano's caldera, the mile-wide sunken amphitheater that was created when the volcano's summit collapsed during some prehistoric eruption. Also, the occasional lava flows and pyroclastic flows—lethal surges of ash and pumice buoyed by hot gases—have generally exited the caldera toward the west, because the caldera's walls have been breached on that side. The city of Pasto, on the eastern side of the volcano, is partially protected by the caldera's 500-foot rampart.

One more feature of Galeras limits the hazard it poses to the local population. Because of its moderate altitude, combined with its location barely eighty miles north of the equator, its summit is free of snow or ice. Snow banks or glaciers may enhance a volcano's beauty, but they also spell danger, because an eruption can rapidly melt the ice. The resulting meltwater, mixed with soil, rock, and ash, is likely to rush downhill in the form of all-consuming mudflows. In 1985, an eruption of the 17,400-foot Nevado del Ruiz volcano, three hundred miles northeast of Galeras, melted its glacial cap: The resulting mudflow traveled more than twenty miles to the town of Armero, which was nearly totally destroyed, at a cost of more than 23,000 lives. Such an event could not happen at Galeras.

Whatever its danger level, Galeras was largely ignored by the world's volcanologists until 1988. In that year, after half a century of inactivity, the volcano showed renewed signs of life. A series of small earthquakes struck the area. In addition, steam began to vent from the volcano. The Colombian government, hypersensitive to volcanic hazards after the Nevado del Ruiz tragedy, sent several volcanologists to investigate and monitor the situation.

Climbing Galeras is a simple matter: One gets into a jeep and drives up. The access road zigzags its way up to the southeastern rim of the caldera. There, in 1988, was located a small police post and several communications towers. A few policemen were always stationed at the post to guard the towers against sabotage by the leftist guerrillas who were active in Nariño province. From the rim, one could look down on the interior of the caldera: Its main feature was a central volcanic cone. About half a mile wide at its base, the cone rose 450 feet from the floor of the caldera but did not rise as high as the caldera rim. At the top of the cone was the actual crater of the volcano—a 100-foot-deep cavity. Getting from the caldera rim to the edge of the crater was an arduous journey: It involved edging down the very steep eastern rampart of the caldera with the aid of a fixed rope, crossing the floor of the caldera (the "moat"), and then climbing the cone. From there, it was another tricky descent into the crater itself.

When the Colombian volcanologists visited the caldera in 1988, they did not descend into the central crater, because it would have meant death to do so: The crater floor was incandescent with heat. Obviously, something very serious was going on inside Galeras. They returned to Pasto and reported their findings to the director of the Colombian National Institute of Geology and Mines, or INGEOMINAS. He called in turn for help from the United States. A few weeks later David Harlow, of the U.S. Geological Survey (USGS) in Menlo Park, California, brought a small team of scientists to Galeras. They installed several more seismographs around the volcano.

In the spring of the 1989, a larger group of Colombian and foreign volcanologists met in Pasto. Two eruptions took place that spring, and though they had both been very small, they got the attention of the people of Pasto. The local government began issuing color-coded warnings, and as so often happens, these induced more confusion than comprehension. Furthermore, the city experienced serious economic problems, as banks stopped issuing loans to local businesses and tourism dried up.

Soon after the 1989 meeting, Galeras quieted down. Then, in the fall of 1991, eruptions began again. A lava dome rose slowly from the floor of the crater, eventually reaching a height of 150 feet. In response to this alarming development, the governor of Nariño called for yet another meeting of scientists, which took place later that month. Among the attendees was Marta Calvache, a young Colombian volcanologist who grew up in the shadow of Galeras. She had been at Nevado del Ruiz during the deadly 1985 eruption. In the aftermath of that event she had met Stanley Williams, an expert in volcanic gases at Louisiana State University. Calvache later went to LSU and did a master's thesis with Williams. (Williams moved to Arizona State University in 1991.) The relationship between Calvache and Williams was an obvious and immediate benefit to Calvache, furthering her expertise and her career. But, as it turned out much later, it was an even greater benefit to Williams, for it was Calvache who saved his life while his colleagues died.

Another attendee at the 1991 meeting was Bernard Chouet, a Swiss-born geophysicist with the USGS in Menlo Park. Chouet's specialty was the interpretation of the seismic signals emitted by active volcanoes. Over the course of a few years before the meeting, Chouet had come to believe that he had discovered a hitherto unknown method for using these signals to predict eruptions.

The seismic signals generated by volcanoes are of two basic kinds. The more common kind are basically little earthquakes: They are produced by the fracturing of rock as magma creates passageways for its ascent to the surface. On seismograms, these events look quite like the common, nonvolcanic earthquakes that are generated by the motion of geological faults: brief, jittery signals that, if sped up and played through a loudspeaker, sound like bangs, pops, rips, crunches, roars, or other unattractive noises. They are assigned magnitudes just like regular earthquakes, and though most are tiny, a few range up to magnitude 5 or so, and are thus easily felt by people living in the vicinity of the volcano. They are called volcano-tectonic earthquakes.

The other, less common seismic signals are quite different: They are low-pitched (infrasonic) vibrations that may continue for half a minute or longer. When sped up and converted into sound, they have an eerily musical quality—they may be reminiscent of whale song, a dirge played on trombones, or Tuvan throat-singing.* Unlike volcano-tectonic earthquakes, these events confine most of their energy to a single, very low frequency—a deep-pitched tone—along with some higher-pitched harmonics that add to the musical quality of the sound. They are called long-period events—"long-period" in this context means the same thing as "low-pitched."

Prior to Chouet's work, much more attention had been paid to the volcano-tectonic earthquakes than to the long-period events as predictors of volcanic eruptions. It is certainly true that most volcanic eruptions are preceded by volcano-tectonic earthquakes. They may start months or years before an eruption, as magma begins to rise from deep magma chambers and collect nearer the surface. These earthquakes are commonly the long-term warning signs that tell volcanologists—and the local populations—that they need to pay attention to the volcano.

Sometimes an intense swarm of volcano-tectonic earthquakes may immediately precede an actual eruption. This happened at the Rabaul caldera in Papua New Guinea in 1994, for example. The port city of Rabaul was evacuated within hours. Next day, the twin volcanoes that guard the harbor entrance both erupted, destroying much of the city. Thanks to the warning provided by the earthquake swarm, only five people died—one of them from a lightning strike.

Yet, just as often, volcano-tectonic earthquakes don't give timely warning of an eruption. An earthquake swarm may occur without a subsequent eruption, or an eruption may occur without a preceding swarm. The fallibility of volcano-tectonic earthquakes as short-term predictors of eruptions is the main reason why local populations are sometimes kept in a prolonged state of needless anxiety or, conversely, given false reassurance in the face of a looming catastrophe.

During the 1980s, Bernard Chouet turned his attention to the

---

*Examples can be heard online at http://www.ees.nmt.edu/Geop/mevo/seismic/tremor.html.

long-period events as possible predictors of eruptions. Chouet had an insight about the nature of these events. He realized that their musicality—the pure tones with their added harmonics—must result from a resonance, that is to say, the reverberation of acoustic waves within a limited space. When Chouet analyzed long-period events mathematically with this idea in mind, he was able to define the characteristics of the system that created them. Long-period events, he concluded, occur when a slug of magma jerks forward within a sheet-like crack in the rock. The motion of the magma is quickly arrested by the confining rock, triggering an acoustic wave. This is analogous to the "water-hammer" that may be generated by household water-pipes when a faucet is turned off too quickly. "It's as if you're pinging it with a hammer blow," Chouet told me during a 2006 interview. "The acoustic pulse will travel through the fluid and hit a boundary and reflect, and keep going back and forth." Nevertheless, some of the acoustic energy escapes into the solid rock and radiates to the surface, where it can be detected by seismic instruments as a long-period event.

According to this model, long-period events will occur when the path of the advancing magma is at least partially obstructed, rather than being open to the surface, just as water-hammer occurs in closed pipes. Obstructed magma is put under increasing pressure as more magma is forced up from the deep. Thus a series of long-period events may signal an increase in pressure that will end only when the pressure within the magma exceeds the weight of the overlying rock. At this point, an explosive eruption will occur.

During the mid-1980s, Chouet looked through the seismographic records of past eruptions at a number of volcanoes. He found long-period events in the run-up to several eruptions, including the disastrous 1985 eruption of Nevado del Ruiz. Then, on January 2, 1991, he successfully predicted an eruption of the Redoubt volcano, which lies one hundred miles southwest of Anchorage, Alaska, and twenty-five miles upriver from an oil-storage terminal. He made the prediction when there was a sudden increase in the frequency of long-period events—they began occurring every minute or so. In response to his prediction, the terminal was closed down and evacuated by five p.m. that same day. Just two hours later the volcano erupted, and the resulting mudflow left parts of the terminal standing in three feet of muddy water.

Thus, when he came down to Pasto for the meeting later that year, Chouet was highly focused on looking for long-period events in the seismographic signals from Galeras. At the time, the volcano was venting excess gases through a long fissure in the lava dome, which opened up every few minutes to release a cloud of gas. Because of the action of this "safety valve" the dome's internal pressure was not building in any dangerous way. Chouet predicted, however, that the crack would eventually seal off, and he advised Calvache and the other Colombian volcanologists to keep a lookout for long-period events once that happened. "When it seals," says Chouet, "the source itself doesn't know it's sealed; it's going to keep on pumping, but now it's pumping into fractures that are embedded in a solid, so instead of pumping into the atmosphere you are pressurizing the whole system."

During the early part of 1992 the lava dome did in fact seal itself off, as minerals deposited by the venting gases clogged the fissure through which they had been emerging. To all external appearances, the volcano was going back into dormancy. But one day in early July, the local volcanologists noticed an odd-looking signal on their seismographs. It was a long-period event, but a particular kind of long-period event in which the vibration quickly built to a peak amplitude and then gradually faded to silence over a period of minutes. On the seismographic recording, the event looked like a screw viewed from the side, and the scientists later named it a *tornillo*—the Spanish for "screw." Eight more *tornillos* occurred over the following five days, and then, on July 16, the lava dome exploded, sending a shower of rocks across the caldera that destroyed some of the communications towers on the caldera rim. It was a small eruption—the city of Pasto was never threatened—but it added strength to Chouet's hypothesis that long-period events could be used to predict eruptions.

Because of Galeras's continued activity, yet another scientific meeting was planned for January of 1993. This time the organizer was Marta Calvache's graduate advisor, Stanley Williams. Williams was a very different kind of volcanologist from Bernard Chouet. For one thing, his specialty was not studying volcanoes' seismic signals, but collecting and analyzing the gases they emit—gases such as water vapor, carbon dioxide, sulfur dioxide, and the rotten-egg gas, hydrogen sulfide. These gases are dissolved in the magma, but as the magma approaches the surface they come out of solution and, if there are open conduits to the surface, they enter the atmosphere at vents called fumaroles. Williams believed that changes in the amount or kinds of gases discharged at fumaroles were potential indicators of impending eruptions. The track record for prediction on this basis was spotty, however.

Williams and Chouet also differed in their styles. Chouet was primarily a theoretical geophysicist, whose work consisted of creating mathematical models on computers at the USGS headquarters in Menlo Park. He did sometimes visit volcanoes, as for example when he went to the 1991 Galeras meeting, but such visits were not really essential to his work. Williams, on the other hand, was very much a field volcanologist. His work required him to enter active volcanoes and approach as close as possible to fumaroles while they were discharging literally tons of superheated, toxic, corrosive, or asphyxiating gases. Thus, even putting aside the chance of an explosive eruption while he was on the volcano, there were significant risks in his work, and he seemed to revel in them. As he acknowledged in his own published accounts, Williams sometimes dispensed with the protective gear that the USGS mandated for their scientists—hard hats, gas masks, fire-resistant coveralls, and the like—and worked in nothing more than street clothes and a stout pair of boots. He spoke of "sucking volcanic gases" as if that were a required initiation rite for would-be volcanologists.

Williams obtained funding for the 1993 meeting from the United Nations, and he was able to invite a stellar cast of experts from around the world. From Russia came Igor Menyailov, like Williams an expert on volcanic gases. From Britain came Geoff Brown, who studied changes in gravity caused by magma movement within a volcano. From Switzerland came Bruno Martinelli, a seismologist and Marta Calvache's long-distance boyfriend. Others came from as far away as Iceland and Japan. The main contingents, however, were from Colombia and the United States. Among the Colombians were three seismologists, Fernando Gil (a collaborator with Chouet in his research on the long-period events), Diego Gomez, and Roberto Torres, as well as two gas experts, José Arles and Nestor Garcia. Marta Calvache also attended; in fact, as a local geologist and Williams's ex-student she did much of the advance planning for the meeting.

The American group, besides Williams, included Fraser Goff, a chemist from the Los Alamos National Laboratory, Andrew Macfarlane of Florida International University, and Charles Wood, a planetary scientist from the University of North Dakota who was working on technologies for predicting volcanic eruptions with satellites. Williams also invited Bernard Chouet as well as other USGS scientists. In what may have been a fateful decision, however, the USGS prohibited its scientists from attending the meeting, citing the unstable security situation in Colombia. Thus Williams and the other meeting participants were deprived of the expertise of the one person who, in retrospect, was best qualified to save them from tragedy. "If I had been down there at the time and I had seen the long-period events," says Chouet, "I would certainly have voiced my opinion that it was not an appropriate time to go into the crater. But I couldn't have just jumped in front of them and said, 'Over my dead body!' so I don't know what the outcome would have been."

The meeting began with three days of scientific talks. Most of the talks were not specifically about Galeras, but two young Colombian seismologists, Diego Gomez and Roberto Torres, did talk (in Spanish) about the seismic history of the volcano, including mention of the *tornillos* that had preceded the eruption of the previous summer. This issue was particularly relevant because after several months of relative calm, the volcano had begun to show signs of renewed activity toward the end of the year. Two days before Christmas a new *tornillo* had been recorded, and further *tornillos* had been occurring at a rate of about one a day, right up through the beginning of the meeting. If Chouet was right, the new series of *tornillos* signaled that the magma was knocking insistently at the roof of the mountain and would soon blow its lava cap to pieces—a scenario that would make a trip to the caldera a risky, even foolhardy, enterprise.

In view of the tragedy that followed, the question of what was discussed about the current status of Galeras has great significance, yet there are very divergent accounts of the matter. One account was presented by geologist-turned-science-writer Victoria Bruce in her 2001 book *No Apparent Danger*. According to Bruce, there was explicit discussion of the *tornillos*, Chouet's theory, and the imminent likelihood of an eruption. The linkage between long-period events and eruptions was, in her eyes, pretty much established scientific fact at the time of the meeting, and Williams decided to go ahead with the field trip in the face of explicit warnings from seismologists that an eruption might be looming. His decision, in Bruce's view, could be explained only by a reckless "cowboy" attitude, perhaps combined with blind belief in the superiority of his own favorite technique—gas analysis.

In his own 2001 book, *Surviving Galeras* (written with Fen Montaigne), Williams painted a very different picture. "As our conference got under way," he wrote:

> INGEOMINAS and foreign seismologists were not alarmed by this desultory pattern of *tornillos;* at the time, such small numbers were considered benign. . . . Only after further eruptions in 1993 did we finally come to understand that small numbers of *tornillos* at Galeras—even as few as one or two a day—*might* presage an eruption. But there was no such understanding then. In the days before our trip into the crater, no one brought the *tornillos* to my attention or warned that the volcano might be poised to blow. . . . Based on all available evidence, the consensus at the observatory was that Galeras was safe.

I asked Charles Wood (who is now at Wheeling Jesuit University in West Virginia) for his view of the matter, and he generally sided with Williams. "My perception is that there was no serious question about whether we should take this field trip," he said. If the Colombian scientists discussed the ominous significance of the *tornillos,* their message was lost on Wood and perhaps the other foreigners too. "My Spanish is rudimentary, and I mostly talked with the English-speakers," he said. "I would say that the work Bernard [Chouet] had done was not well enough known, and probably not well enough documented—I'm speculating—to say that if you see any of these screw-type features, you know there's going to be an eruption in the next twenty-four hours or whatever. It's not clear to me that there was that level of aware-ness of their predictive value." Wood also complained that Vir-ginia Bruce, who interviewed him twice for her book, was "not honest" in her dealings with him: She never disclosed her critical views of Williams, he said, and thus gave Wood no motivation or opportunity to defend him.

Another scientist who took issue with Bruce's book was Chouet's colleague, Fernando Gil. In Bruce's account, Gil had a conversation with Williams on the night before the field trip, in which he warned Williams about the *tornillos* and emphasized the danger of an impend-ing eruption. Williams, Bruce says, refused to cancel the trip, though he did agree to cut down the number of scientists who would be al-lowed to enter the crater. But when Bruce's book was published in 1991, Gil told the *Chronicle of Higher Education* that this purported conversation between him and Williams didn't happen.

The field trip was originally scheduled for the third day of the meet-ing, January 13. In what would turn out to be a fateful change of plan, however, the organizers rescheduled the trip for the following day—the reason being a planned power outage on January 14 that would make indoor lectures impossible on that day. According to Wood, heavy rain on the thirteenth was another factor in the deci-sion to postpone the trip. In any event, the scientists set out early on the morning of the fourteenth in a fleet of vehicles, and they reached the rim of the caldera at about nine a.m.

It had stopped raining, but thick clouds and fog enveloped the volcano's summit, reducing visibility to a few feet. The interior of the caldera was completely hidden. "It was a disappointing touristic experience," said Wood. "It was cold; we were all in our down jackets, milling around, getting some talks from different people."

By ten a.m. the fog had lifted slightly—enough, in Williams's judgment, to permit the descent into the caldera. Only twelve scientists were to accompany Williams on that leg of the field trip, however. As had been previously arranged, the remaining scientists were to make a variety of other trips around the flanks of the volcano. Wood, for example, joined a group led by Marta Calvache and Patty Mothes, a young American volcanologist living in Ecuador. Their group would explore some deposits left by earlier eruptions, about a half-mile below the caldera rim. Wood was quite happy not to join Williams's party. "Just going down into the caldera with that poor visibility would be dangerous," he said.

One by one, the members of Williams's group backed down the steep wall of the caldera while clinging to the fixed rope. Because of the fog, they were quickly lost to view from the rim. Within the caldera, however, visibility was better. By the time the scientists reached the "moat"—the lowest part of the caldera between the outer ramparts and the central volcanic cone—they could see one another and the cone ahead of them. After crossing the moat they began trudging up the side of the cone toward the crater rim. It was slow going; the terrain was steep and exceedingly rough on account of the rocks and cinders that had been scattered across the slope during previous eruptions. The altitude of more than 13,000 feet didn't help either: Some of the party were decidedly out of shape and needed to pause for breath after every few steps.

Among the thirteen men, only one was wearing full protective gear. This was Andy Adams of Los Alamos: As a U.S. government employee he had to follow safety guidelines that mandated a hard hat, steel-toed boots, and fire-resistant coveralls. One other scientist, the Guatamalan Alfredo Roldan, wore a hard hat. Most of the others were bare-headed and wore down parkas. Williams poked fun at

Adams for his seemingly excessive safety-consciousness, something that Adams didn't find amusing.

Three hikers followed the scientists into the caldera. These were a local professor, his teenage son, and his son's friend, who were visiting the caldera out of pure curiosity. They were dressed even less suitably than Williams: Rather than boots they were wearing sneakers—a poor choice for navigating the clinker-strewn slopes of the volcanic cone.

Once the scientists reached the top of the cone they could look down into the crater—a depression about 400 feet wide, surrounded by almost sheer 100-foot-high walls. Clouds of gas and steam were emerging from various spots on the crater floor, as well as from a group of fumaroles named Deformes on the southwest part of the crater rim. Aside from the gas venting, which had been going on for months, there was no sign of volcanic activity in the crater.

About this time, the seismologist who was watching the seismographs back at the Observatory in Pasto noted the occurrence of a *tornillo*. She radioed the information up to Roberto Torres, who was on the caldera rim, and he relayed it to José Arles on the cone. José noted the information but didn't consider it particularly worrisome—fifteen *tornillos* had occurred over the previous three weeks, after all, without the volcano having shown the slightest sign of erupting.

The scientists in the caldera now split up into small groups to carry out a variety of tasks and observations. Geoff Brown, the British gravity expert, led a group around the rim of the crater. They took measurements of gravity as they went, using a bulky but highly sensitive piece of equipment that they had lugged up the cone. Several other members of the group, including the chemist Andrew Macfarlane, gathered at the Deformes fumaroles to measure the temperature of the emitted gases and to collect samples for later analysis in the laboratory. The various tasks kept the scientists busy for a couple of hours.

Around noon, two members of the party climbed down into the crater itself. These were two chemists, the Russian Igor Menyailov and the Colombian Nestor Garcia. They had finished sampling at the Deformes fumaroles and wanted to take more samples at the vents within the crater. The other scientists, including Williams, were content to watch from the crater rim. Williams took the opportunity to chat with the three hikers. About an hour later, most of the scientists began their return trip to the rim of the caldera—a trip that involved the descent of the central cone, the crossing of the moat, and the strenuous ascent of the 400-foot-high caldera wall.

At 1:41 p.m., the members of the party were positioned thus: one Colombian, Alfredo Roldan, had just completed the journey and was standing on the caldera rim giving an interview to reporters from a local television station. Two other Colombians, Fabio Garcia and Carlos Estrada, were in the process of climbing the caldera wall with the aid of the fixed rope. Andy Adams, the safety-conscious Los Alamos employee, was at the base of the caldera wall, preparing to start the climb. Eight other men—Williams, Macfarlane, Arles, an Ecuadoran geochemist named Luis LeMarie, an American geochemist named Mike Conway, and the three local hikers—were at various points between the top and the base of the central cone. Geoff Brown and the Colombians Carlos Trujillo and Fernando Cuenca were standing on the opposite, western side of the crater rim, far from the trail that led out of the caldera. Igor Menyailov and Nestor Garcia were still on the floor of the crater.

Precisely at 1:41 p.m. was the moment when the upward force exerted by the magma within the volcano's cone finally exceeded the downward force exerted by the weight of the overlying rock, and a devastating explosion ensued. The blast was followed by an unending roar as rocks and magma flew skyward. The blast ripped the floor of the crater to shreds, instantly killing Igor Menyailov and Nestor Garcia and vaporizing their bodies. The three men who had been standing on the western rim of the crater were also killed instantly. Carlos Trujillo's body was cut cleanly in half by a flying boulder; as for Geoff Brown and Fernando Cuenca, nothing was ever found of them beyond small fragments of flesh stuck to rocks.

The eight men who were descending the cone broke into an instinctive mad downward rush. A barrage of incandescent rocks—some as small as bullets, some as big as television sets—began falling among them. As the rocks struck the ground they exploded like shells, firing off secondary showers of red-hot shrapnel in all directions.

Four of the men—José Arles and all three of the local hikers—were killed within seconds as flying rocks stove in their skulls and smashed their bodies. Andrew Macfarlane was knocked down several times, once by a rock that struck his forehead, creating a hairline fracture in his skull and sending blood pouring into one eye. For a brief time he took refuge behind a rock, then he continued on his stumbling downward run. Every time he fell, his hands or body were seared by the rocks they touched, and he was forced to get up and struggle on. Then he came across Mike Conway and Luis LeMarie, who had found partial protection in a small hollow. The three men hid there for about five minutes, while the rain of rocks gradually abated and changed to harmless ash.

Stanley Williams also took off running at the first explosion, but he was quickly knocked down by falling rocks. One rock smashed his right lower leg with such force as to almost completely sever it: His foot was left dangling at an unnatural angle from the leg, connected to it by a few strips of flesh. His other leg was also broken, and a third rock struck his head above his left ear: It dislocated his jaw, partially detached the retinas of both eyes, and drove fragments of skull into his brain. Smaller rocks burned holes in the skin of his back. He lay helpless, just a few feet from the lifeless body of José Arles. In spite of his terrible injuries, Williams remained alive and conscious, and he cried out for help. For a long time he received none.

Roldan's television interview on the rim of the caldera was rudely interrupted by the blast, as both he and the newsmen dived for cover. Roldan was struck on the head by a flying rock, but his hard hat saved his life. He crouched down behind the rim of the caldera and watched a cloud of ash and hot gas rise from the crater and reach more than a mile into the sky. At about the same time one of the policemen who was stationed at the guardhouse was also struck: The rock hit his lower arm and completely amputated one of his hands.

Fabio Garcia and Carlos Estrada, who had been climbing the caldera wall at the time of the blast, were able to complete their climb safely, though they had to dodge a hail of rocks during the first five minutes. Andy Adams, who had not yet begun the climb at the time of the blast, was also able to climb out unaided. He was struck by innumerable small rocks, but his fireproof clothing and hard hat saved him from serious injury—he sustained only some burns on his unprotected neck.

After the initial barrage of rocks subsided, Mike Conway, Luis LeMarie, and Andy Macfarlane left their foxhole and began to make their way down the cone and out of the caldera. It was slow going. All three men had numerous cuts and burns. LeMarie had fractures in both legs. Macfarlane had several severe lacerations in addition to his head wound. It wasn't until an hour after the eruption that the three men made it down to the moat, and then they faced the arduous ascent of the caldera wall. Conway, the least injured, was able to clamber to the caldera rim unaided. LeMarie was assisted up the rope by Carlos Estrada, who had climbed back down into the caldera to help him. Macfarlane could only make it as far as the bottom of the rope before he collapsed in pain and exhaustion.

The initial explosion had alerted the groups of scientists who were on the outer flanks of the volcano. From where Marta Calvache and Patty Mothes's group was positioned, about halfway down the volcano, the sound was not especially loud. "We were having lunch," said Wood. "We looked up, thinking, 'What was that?' and we saw the [ash] cloud. Marta and Patty said, 'OK, we're going up; we have the most experience with this volcano—the rest of you evacuate down to Pasto. So they got into a jeep and went up to the summit."

It took Calvache and Mothes about thirty minutes to reach the summit, because they first had to hike from their location to the access road where their vehicle was parked. Then, on the way up the road they met a military truck that contained most of the policemen who had been stationed at the guardhouse, including the man who had lost a hand. Calvache told a couple of the men to accompany them back to the summit in order to help with the rescue.

When they got to the summit, they were met by a chaotic scene: a battered guardhouse, vehicles with windows smashed out, blood and rocks on the ground, and a large number of people, including a civilian rescue squad, who were milling about ineffectually. No one had dared to descend into the caldera, but at that same moment two Colombians from the Observatory in Pasto, Ricardo Villota and Milton Ordoñez, arrived on the scene. They had been alerted by the violent signals that were appearing on their seismographs and, fearing the worst, they had rushed up the mountain in a truck. Villota and Ordoñez immediately began descending the steep rampart, and Calvache and Mothes followed them. They could see Andrew Macfarlane lying half-conscious at the bottom of the fixed rope, and although they couldn't see Stanley Williams, they could hear his calls for help. Thus they were spurred on to attempt a rescue, in spite of the danger of a renewed eruption.

When they reached Macfarlane, Ordoñez and Mothes started to attend to his wounds. Soon, other helpers arrived, and eventually they were able to get Macfarlane onto a stretcher, haul him up to the caldera rim, and place him in an ambulance that rushed him off to a hospital in Pasto.

Meanwhile Calvache and Villota had gone off in search of other survivors. They soon saw that the three local hikers were dead. José Arles was dead too, with a huge gash in his skull that exposed most of his brain, but Villota, a close friend of Arles, made a futile attempt to revive him by raising him into a sitting position, whereupon his brain fell out onto the rocks. Villota was too shocked to continue with the rescue efforts.

Stanley Williams was the only person still alive on the volcanic cone, but he was horribly wounded, with a nearly severed right foot, a broken left leg, a depressed fracture of his skull, a broken jaw, and burns and lacerations over much of his body. Calvache, Mothes, Ordoñez, and another rescuer placed Williams in a blanket and, painfully slowly, hauled him down the cone and across the moat. The journey of a few hundred yards took nearly two hours, and throughout that time the volcano continued its ominous rumblings. By the time they arrived at the base of the caldera wall, professional rescuers had reached the scene. Williams was placed on a stretcher, hauled up to the rim of the caldera, and airlifted by helicopter to a hospital in Pasto. Calvache and Mothes continued their search for more survivors, but by six p.m. they realized there were none, and they in turn were airlifted off the mountain.

Williams had been rescued, but his ordeal had barely begun. At a hospital in Pasto, he had surgery to remove a blood clot and bone fragments from his brain. His legs were put in plaster casts, and he was flown by air ambulance to Bogotá and from Bogotá back to the United States. There he underwent further surgery to his head and legs: After some unsuccessful operations to repair his shattered right leg, the entire lower leg was stabilized with a "birdcage" of embedded metal rods, which Williams wore for eleven months. He also had operations—only partially successful—to restore the hearing in his left ear. Later, Williams developed pneumonia, and he also had a grand mal epileptic seizure.

Besides the physical problems, Williams was plagued by other kinds of difficulties. Whether on account of the horrifying experience he had lived through or because of the brain damage he had suffered, Williams recalled events incorrectly. In media interviews—of which he gave many—he allowed himself to be presented as the sole survivor of the scientists who had been on the volcanic cone, or even of the entire party who entered the caldera. This naturally irked the other survivors, who came to see Williams as a publicity hound.

Allegations that Williams had ignored the signs of an impending eruption culminated in Victoria Bruce's scathing book. Some of Williams's colleagues echoed Bruce's point of view. In a 2001 review in *Science,* volcanologist Haraldur Sigurdsson of the University of Rhode Island took issue with Williams's claim that "the best work . . . comes from those of us who walk into the crater." Dismissing this as more bravado than science, Sigurdsson asserted that the most modern techniques in volcanology don't require risky excursions into danger zones. Still, there have been other volcanologists, such as Charles Wood, who have steadfastly defended Williams.

In response to the Galeras tragedy, as well as a string of other events that had taken the lives of volcanologists, a group of scientists put together a set of safety guidelines that was published in 1994 by the International Association of Volcanology and Chemistry of the Earth's Interior. The guidelines recommend that scientists should not visit active volcanoes unless it is absolutely necessary for data gathering: Field trips to active volcanoes should not be offered as add-ons to scientific meetings. The guidelines also describe the kinds of protective gear that should be worn, as well as other precautions that should be taken, such as radio communications, emergency preparedness, and so on.

But the guidelines haven't put a stop to the deaths and injuries, because volcanologists often disregard them. In July 2000, for example, an international group of scientists climbed Mt. Semeru, a volcano on the island of Java. The trip, which followed a volcanological meeting, had no particular scientific purpose, and there had been no plan to approach the active crater, but some of the scientists did so on the spur of the moment, without any kind of protective gear or other safety precautions. A small explosive eruption, similar to the one that killed Stanley Williams's colleagues on Galeras, happened without warning while the scientists were standing on the edge of the crater. It killed two Indonesians and injured three Americans, one of them severely.

While this accident may have been the result of mere thoughtlessness, a culture of daredevilry still permeates the volcanological community. To take one example, the Web site of John Stix, one of the organizers of the 1993 Galeras meeting, shows a student posing *Jackass*-like in front of a spray of red-hot lava, while wearing no more protection than a cotton cap. It is as if the people who are drawn to do research on active volcanoes are exactly the ones who are psychologically ill-suited for the job.

Stanley Williams has remained professionally active, but both he and others have noted a sharp decline in his intellectual powers and productivity, which he attributes to his brain injury. Williams has had other travails in recent years, including a divorce and a bizarre episode in 2001 when he was briefly suspected (but quickly cleared) of involvement in a double homicide—an event that brought him another deluge of publicity.

Whatever the views of others, Williams himself (who didn't respond to my requests for an interview) has steadfastly refused to accept any blame for the Galeras tragedy. "I do not feel guilty about the deaths of my colleagues," he wrote in his book. "There is no guilt. There was only an eruption."

CHAPTER 4

NEUROSCIENCE:
The Ecstasy and the Agony

In 2002, when a neurologist at Johns Hopkins School of Medicine in Baltimore published a study on the toxic effects of the drug Ecstasy, his findings bolstered a political campaign against recreational use of the drug. Only much later did it turn out that the findings were the consequence of an almost laughable laboratory blunder.

The year 2002 was a high-profile one for Ecstasy. On June 18, Joseph Biden, the senior Democratic senator from Delaware, introduced the Reducing Americans' Vulnerability to Ecstasy Act into the U.S. Senate. As its acronym suggested, the RAVE Act was directed against the scenes where the use of Ecstasy and other "club drugs" was most in evidence: the all-night music and dance parties known as raves. (Ecstasy use has since expanded to other venues, such as college campuses.)

Although Biden is best known as a foreign-policy expert, he also has a long history of involvement in drug-control legislation, including the laws that created the Office of National Drug Control Policy (the "Drug Czar"). The RAVE Act was actually an amendment to a section of the existing Controlled Substances Act—the so-called crack-house statute—which allowed prosecutors to seek the destruction of premises used for drug sales or drug use. Biden's bill allowed for $250,000 fines against persons who promoted or provided space for raves if they knowingly permitted the use of illegal drugs such as Ecstasy at these events.

In the preamble of the bill, Biden depicted rave organizers as deliberately fostering a drug culture under the hypocritical guise of concern for partygoers' welfare. "Because rave promoters know that Ecstasy causes the body temperature in a user to rise and as a result causes the user to become very thirsty," Biden wrote, "many rave promoters facilitate and profit from flagrant drug use at rave parties or events by selling overpriced bottles of water and charging entrance fees to 'chill-rooms' where users can cool down." Seemingly well-meaning security measures were fraught with malevolent intent. "Some [rave promoters] even go so far as to hire off-duty, uniformed police officers to patrol outside of the venue to give parents the impression that the event is safe," according to Biden's bill.

The RAVE Act represented a certain change of emphasis in the war against drugs. With the traditional targets—people who used or dealt in hard drugs like heroin and crack cocaine—there was little social controversy about the value of drug-control measures. With those drugs, the public perception was of random acts of violence committed by addicts in a desperate quest to finance their next fix, or by warring gang lords in disputes over their drug fiefdoms. With club drugs like Ecstasy, on the other hand, a threat to public order was much less apparent. For most users, Ecstasy is not an addictive drug—not in the traditional sense of generating an all-consuming physical dependency, at least. Because Ecstasy is generally used on an occasional basis, it doesn't usually threaten its users with financial ruin. Nor, in most cases, does it threaten their jobs, studies, or relationships. Ecstasy use, in other words, is largely compatible with a conventional middle-class lifestyle.

Lacking a public-order platform, the advocates of strict measures against Ecstasy concentrated on the dangers that the drug might pose to its users' health. These dangers certainly exist. Most well-recognized is the risk of acute hyperthermia—heatstroke—while under the influence of the drug. Every year, a few Ecstasy users die of hyperthermia or of hyponatremia (overdilution of the blood caused by drinking excessive amounts of water to keep cool). Ecstasy users sometimes die in traffic accidents while under the influence of the drug, or from other drugs that are taken along with Ecstasy or that are present as adulterants in Ecstasy tablets.

Still, Ecstasy causes only a handful of acute deaths in the United States per year, and relatively few acute health problems of any kind. In the year before Senator Biden introduced the RAVE Act, fewer than 2 persons in 100,000 visited an emergency room with an Ecstasy-related problem, compared with 76 per 100,000 with a cocaine-related problem, according to the Drug Abuse Warning Network.

Although the acute risks are not especially great, there are some indications of long-term health hazards associated with Ecstasy use. For some users, the initial positive feelings engendered by the drug—euphoria, energy, intensified sensations, and a deep sense of intimacy with others—fade with repeated use, to be replaced by depression, memory difficulties, sleep disorders, and cognitive problems. The prevalence of these ill-effects is a topic of debate: Many of the people who experience them had taken a variety of drugs, making the specific role of Ecstasy difficult to tease out. And conceivably, cause and effect are at least partly the other way around—a disposition to these mental problems might promote drug use, rather than drug use triggering the problems.

Given the nebulous nature of some of Ecstasy's reported ill-effects, educating young people to avoid the drug has always been a challenge. One strategy has been to replace the reported psychological hazards with a very concrete physical effect—damage to the brain.

Already in 1987, the U.S. government's National Youth Anti-Drug Media Campaign tried this approach with a series of TV commercials that featured a chicken egg ("This is your brain") and the same egg frying in a skillet ("This is your brain on drugs"). For some reason, the commercials were extraordinarily memorable, even iconic—they spawned endless spin-offs, imitations, and parodies. Whether they achieved their goal of reducing drug use was less clear. "When I saw people that were on the high school honor roll smoking pot, I realized that the commercial's message was false," one young woman told CNN. "I remember thinking, 'When are their brains going to fry?'" By 2003, more than one billion federal dollars had been spent on this and related media campaigns, but they had no significant effect on drug use among young people, according to a U.S. government study.

In the case of Ecstasy, the main evidence for brain damage was provided by George Ricaurte, M.D., Ph.D., a neurologist at the Johns Hopkins University School of Medicine. Ricaurte (pronounced ri-CART-ey) had made a name for himself by investigating the damaging effects of several illegal drugs on the brain, and he lost no opportunity to stress the significance of his findings to potential drug users. As such, he was a darling of the National Institute on Drug Abuse (NIDA). Along with psychiatrist Una McCann, who is his wife and collaborator, Ricaurte received nearly $16 million in funding from NIDA and other government agencies between 1989 and 2003.

In 1998 Ricaurte, McCann, and three junior colleagues had published a study that purported to show brain damage resulting from recreational use of Ecstasy. Ecstasy is a chemical derivative of amphetamine: Its scientific name is 3,4 methylenedioxymethamphetamine, or MDMA. It causes a surgelike release of the "feel-good" neurotransmitter serotonin, as well as a more modest release of another transmitter, dopamine. In the 1998 study, Ricaurte's group used a radioactive tracer, combined with positron-emission tomography (PET) scanning, to visualize the distribution of a serotonin-related molecule in the brains of persons who had used Ecstasy in the past, as well as in control subjects who had not. The resulting images showed less of the serotonin tracer in the Ecstasy users than in the controls, and some patches of brain seemed to be devoid of the tracer altogether.

Ricaurte described these results as evidence of structural changes in the brain induced by Ecstasy use—a bit of a semantic stretch, perhaps, when referring to the distribution of a molecule. His paper did include some cautionary statements, such an admission that the changes might not be permanent. And in fact, a German research group, who used similar techniques to study Ecstasy users at various lengths of time after their last use of the drug, has recently reported that the effects of Ecstasy ingestion on the brain's serotonin system do at least partially reverse themselves with time.

NIDA administrators figured that they could illustrate Ricuarte's message without harming any eggs: They printed up and distributed thousands of postcards and posters carrying images taken directly from Ricaurte's paper. These images (which can still easily be found on the Internet) juxtaposed a PET scan of the left side of the brain of a control subject with the right side of the brain of an Ecstasy user. Absurdly, most of the front of the drug user's brain appeared to be missing, and the rest of it was riddled with large holes. By presenting PET scans as if they were actual pictures of brains, the poster heightened the paradox raised by CNN's interviewee. Forget about honor rolls—how could a person even live and breathe with such a horrendous injury?

When I interviewed Ricaurte in 2006, I was struck by the contrast with Robert Iacono, the maverick neurosurgeon I interviewed for chapter 1. Whereas Iacono was by turns humorous, pithy, and abrasive—in other words, an interviewer's dream—Ricaurte was drily circumspect, rarely failing to discuss every possible side of a question in his desire for accuracy and fairness. When I asked him whether he thought the poster campaign was misleading, for example, he said, "An investigator does not play any active part in coming up with these things—the campaigns that are taken on. I don't even know if it's appropriate to label it a campaign. I certainly would endorse the research that generates findings that are relevant to the community or public health. I think it would be irresponsible to not make those results available to the public that pays for the research. If we're doing research in the lab, the people who pay for that research should be privy to the results of that work. As you know, communicating results of scientific studies to the lay public is an art in itself; it's not always easy to speak in laymen's terms. Having said that, I think it becomes exceedingly important to make sure that the data that are being presented communicate the message in as accurate and responsible a way as possible. So the images you refer to, if what they are conveying is the impression that MDMA produces holes in people's brains, somebody should have caught that and said, wait a minute, that is not what MDMA is doing; we think based on this research that it's damaging serotonin neurons. And how to convey that notion appropriately with the appropriate recognitions of animal data and dosages and how much of the animal data generalizes to humans—

there are a number of limitations that should be recognized. If you go to the article where those images come from, we try to explain very clearly what the images are showing." '

Although Ricaurte has generally been well-regarded within the academic community, he has had his critics, especially among researchers who thought that Ecstasy might have therapeutic applications. One such person was Charles Grob, director of child and adolescent psychiatry at Harbor-UCLA Medical Center in Southern California. As he told me in a 2006 interview, Grob believes that Ecstasy, given in a carefully controlled clinical psychotherapeutic setting, might have value in the treatment of posttraumatic stress disorder, but he doesn't support legalization of Ecstasy for general, recreational use. A year before Senator Biden introduced the RAVE Act, Grob testified before the U.S. Sentencing Commission. His remarks included the following:

> Carefully examining the record of human research with MDMA, particularly the NIDA-funded studies of George Ricaurte, one observes a persistent pattern of poorly controlled studies, often with deliberate exclusion of vital data sets from published reports as well as unreported pre-selection biases in criteria used to recruit research subjects, which have led to grossly exaggerated and misleading claims in the scientific literature and in the media.

Unperturbed by such criticism, Ricaurte pushed his Ecstasy research in a new direction. Earlier reports in the scientific literature, based on experiments in rodents, suggested that Ecstasy might also have damaging effects on neurons that used dopamine—the other transmitter whose release is triggered by Ecstasy. The doses needed to produce the toxic dopamine effects were rather high, and the damage didn't appear to be as great as the damage suffered by the serotonin system. In 2000, however, Ricaurte set out to establish whether Ecstasy, even when given in a dose comparable to what a person might consume in the course of a single rave, could damage dopamine neurons in the brains of monkeys.

Ricaurte gave pure MDMA (Ecstasy) to five squirrel monkeys and five baboons. Most of the animals received three doses of the drug, spaced at three-hour intervals. The total dose was 6 milligrams per kilogram of body weight. This would correspond to a person taking three pure Ecstasy tablets of about 150 mg each in the course of a single night's partying—an amount that is within the range of what some rave attendees might consume, though probably higher than the typical user's consumption. Because dosing monkeys orally is inconvenient—monkeys often reject food that has been spiked with drugs—Ricaurte gave the animals their MDMA doses by subcutaneous injection.

Two of the animals—one squirrel monkey and one baboon—developed uncontrollable hyperthermia soon after their final dose, and they died within hours. Another two monkeys—again, one of each species—fell ill after their second doses, and were therefore not given their final dose. The four surviving animals of each species were allowed to live for two to eight weeks, and then killed so that their brains could be studied, along with those of several control animals who had received injections of saltwater.

The results were striking: The monkeys that had received MDMA, even those who only received two of the planned three injections, showed signs of profound damage to the dopamine systems in the brain. Contrary to what might have been expected on the basis of previous studies, the damage to the dopamine system was even more severe than the damage to the serotonin system. And it wasn't just a reduction in the levels of dopamine-related molecules, though that had certainly happened. In addition, Ricaurte found that the nerve endings of the dopamine neurons were physically degenerating. The cell bodies may have survived, but their terminals—the all-important sites of transmitter release—were shriveling and dying like autumn leaves. And in response to the destruction, a special set of inflammatory cells were enlarging and multiplying in the affected areas of the brain. It seemed like the kind of thing that could give an Ecstasy-abusing partygoer the morning after from hell.

As has already been discussed in the first chapter of this book, the loss of brain dopamine function is a central feature of Parkinson's disease, the movement disorder that strikes about 60,000 Americans every year and that ends up killing many of them. So the finding of severe damage to the dopamine system in his monkeys' brains immediately provoked an alarming thought in Ricaurte's mind: Could a person who indulged in a single night's use of Ecstasy be setting themselves up for a lifetime of Parkinson's disease?

There was a precedent. In 1976 a Maryland college student by the name of Barry Kidston cooked up a novel designer drug, mainlined it, and went into a state of permanently suspended animation—he simply couldn't move or speak. His condition resembled the most advanced stage of untreated Parkinson's disease. Kidston eventually responded to treatment with L-dopa, the standard therapy for Parkinson's disease, but he later died of a cocaine overdose on the campus of the National Institutes of Health. When NIH researchers examined his brain they found that most of the dopamine neurons in his substantia nigra had died. A few years later, six drug abusers in California were similarly affected.

The evidence that Ecstasy use might cause a Parkinson-like condition alarmed Ricaurte, but it was also an opportunity for public education. In September 2002, Ricaurte and his colleagues published their results in *Science,* under the title "Severe dopaminergic neurotoxicity in primates after a common recreational dose regimen of MDMA ('Ecstasy')." After presenting their data, Ricaurte's group concluded as follows:

> These findings suggest that humans who use repeated doses of MDMA over several hours are at high risk for incurring severe brain dopaminergic neural injury (along with significant serotonergic neurotoxicity). This injury, together with the decline in dopaminergic function known to occur with age, may put these individuals at increased risk for developing Parkinsonism and other neuropsychiatric diseases involving brain dopamine/serotonin deficiency, either as young adults or later in life.

There was only one problem with this line of thought: There were no reports of anyone having put themselves into a "frozen" state by using Ecstasy, whether for one night or over a lifetime. In fact, there wasn't any documented relationship between Ecstasy use and Parkinson's disease or other disorders of movement.

According to Ricaurte, however, this could simply be the result of a failure to look for a connection. When young people fell ill with Parkinson's disease, he wrote, doctors didn't usually inquire about their past use of recreational drugs, so the causative role of Ecstasy might have been missed. Furthermore, it was known that a large fraction of a person's dopamine neurons—about 80 percent of them—have to die before the symptoms of Parkinson's disease become apparent. Maybe, Ricaurte suggested, Ecstasy users didn't destroy a large enough portion of their dopamine system to cause symptoms immediately, but they would nevertheless develop the disorder years or decades later, as natural attrition finished off the job that drug abuse had begun.

Ricaurte's study was a shot in the arm for the RAVE Act. Joe Biden's bill had faltered since its introduction three months earlier. It had sparked numerous demonstrations and protests, as well as thousands of letters, from people who saw the bill as threatening an innocent and popular pastime—raves. What sane person would organize a rave or allow their premises to be used for one, it was asked, if they risked a quarter-million-dollar fine every time a participant popped a pill? Organized opposition to the bill also came from the electronic music industry. In response to the protests, two of the bill's cosponsors withdrew their sponsorship.

When Ricaurte's study was published, public statements from Ricaurte, Johns Hopkins Medical School, NIDA, *Science*, and other authoritative sources painted Ecstasy as a proven threat to human health. Within weeks, Ricaurte's study was being cited in Congress. In October of 2002 Asa Hutchinson, director of the Drug Enforcement Agency (DEA), told a House Judiciary Subcommittee that "[Ricaurte's] study discovered evidence that severe brain damage occurs to the nerve cells which produce the neurotransmitter dopamine in the area of the brain controlling movement. The study concluded that neurological damage could stay hidden for years and increase the risk of Parkinson's disease and associated movement-related disorders." According to Charles Grob, Ricaurte's work "created an atmosphere of hysteria" that fostered this and other legislation related to Ecstasy.

The RAVE Act didn't pass in 2002, but early in 2003 Biden reintroduced it under a new name—the Illicit Drug Anti-Proliferation Act. Biden added it as an amendment to the Amber Alert Act, whose purpose was to facilitate the rescue of abducted children. Riding on those popular coattails, the Act cruised through both Houses without resistance or even discussion, and on April 30, 2003, it was signed into law by President Bush.

Ricaurte's paper naturally incurred the scorn of the rave and drug-liberalization communities. Within academe it was mostly well received, but it did have some detractors—mostly the people, such as Charles Grob, who wanted to legalize Ecstasy as a therapeutic drug. Grob told me that he was "incredulous" when he read the paper. Another critic in this group was Rick Doblin, Ph.D., a public-health specialist and founder of the Multidisciplinary Association for Psychedelic Studies (MAPS). Doblin had long sought to sponsor a trial of Ecstasy, in conjunction with psychotherapy, for the treatment of posttraumatic stress disorder. Doblin is a more radical figure than Grob in that, as he told me quite frankly in 2006, he sees legalizing Ecstasy for therapeutic purposes to be an initial step toward a long-term goal of a general decriminalization of the drug.

Doblin had teamed up with Michael Mithoefer, a South Carolina psychiatrist, who would conduct the actual study. After many setbacks over a period of years, it seemed in 2002 that their project was about to be approved by the Food and Drug Administration. But, according to Doblin, Ricaurte's wife and collaborator, Una McCann, wrote to the Institutional Review Board (IRB) that was considering the application and spelled out what she considered to be the neurotoxic effects of MDMA, including its newly discovered damaging effects on the dopamine system.

In a 2007 e-mail, McCann gave me a different account of the

interaction. She said that it took the form of a brief telephone conversation initiated by one of the IRB members. In the conversation, according to McCann, she emphasized the potential for damage to the serotonin system rather than to the dopamine system, and she said that she would have no problem with the proposed study so long as the subjects were made aware of the serotonin toxicity and the study was conducted in a safe environment.

Whether McCann had any role in the matter or not, the IRB dropped its support of the application, leaving Doblin and Mithoefer high and dry.

Shortly after Ricaurte's paper was published, Doblin and Mithoefer wrote a critical letter to *Science*. According to the letter, which wasn't published until the following June, there were four main reasons for believing that Ricaurte's findings were incorrect or at least irrelevant to the human use of Ecstasy. First, the fact that two of his ten monkeys died, and two fell acutely ill, suggested that the dose of MDMA they were given didn't correspond to a typical human dose, given that the vast majority of Ecstasy users suffered no acute ill-effects. In fact, Ricaurte himself in an earlier study had reported that MDMA given by injection was twice as potent as when given by mouth, suggesting that he had effectively overdosed his monkeys. Second, the letter pointed out that several previous primate and human studies, including some by Ricaurte himself, failed to find damage to the dopamine system from Ecstasy use. Thirdly—in a comment that later turned out to be uncannily perspicacious—it pointed out that the pattern of damage described in Ricaurte's paper resembled the damage that was known to be produced by administration of high doses of methamphetamine (speed), yet that drug did not produce Parkinson-like symptoms in humans. Finally, it reminded readers that there was no reported association between Ecstasy use and Parkinson's disease.

Ricaurte responded to the letter in a somewhat awkward fashion. Only one monkey had died, he said—contradicting what he had written in the published paper. (When *Science* requested clarification on this, Ricaurte explained that he had been talking about the squirrel monkeys—as if he had temporarily forgotten that baboons are monkeys too.) As to his earlier study that reported a difference between oral and injected dosages, Ricaurte responded by citing other studies that contradicted his own. The reason other studies didn't find effects of MDMA on the dopamine system, Ricaurte suggested, was that they didn't employ the three-in-a-row dosage regime that was designed to mimic a one-night Ecstasy binge. And as for the effects of methamphetamine, Ricaurte cited studies that *did* find some indication of Parkinson-like effects. Ricaurte wrapped up his response with a none-too-subtle dig at the motive behind Doblin and Mithoefer's letter: He argued that clinical trials of Ecstasy should not be permitted because of the possible health risks that his research had demonstrated.

Although Ricaurte seemed unruffled by Doblin's attacks, they did in fact motivate him to undertake new experiments. Specifically, he decided to repeat the published experiments, but administering the MDMA by mouth rather than by injection. In this way he would circumvent the comparable-dosage issue and would approximate more closely the actual drug experience of an Ecstasy-using partygoer. Starting just a month after his *Science* paper was published, Ricaurte treated a series of monkeys with the same three-dose regime of MDMA, but he gave the drug by mouth and varied the total dose to levels both higher and lower than those that he had given in the published paper. Although the animals did show signs of damage to their serotonin system, Ricaurte was surprised to find that none of them exhibited damage to their dopamine system. In order to be able to directly compare the oral and injection routes, Ricaurte treated another group of monkeys with Ecstasy by injection: In other words, he repeated his published study. None of these monkeys showed any impairment of their dopamine system either.

One can only imagine Ricaurte's state of mind at this point. Not being able to reproduce one's own published (and much publicized) findings has to be a scientist's worst nightmare. To his credit, he reacted just as he should have done. Rather than stonewalling, ignoring the new findings, or fleeing the country, he set out very systematically to identify the source of the discrepancy between the results of the two studies.

Ricaurte first thought that the failure to replicate his earlier results might have to do with the temperature at which the monkeys were housed during the experiment. This was because high temperatures were known to worsen the effects of Ecstasy. So, in March and April of 2003, he injected another group of squirrel monkeys with MDMA and housed some of them at the normal animal-house temperature and some at a higher temperature. Again, neither group of monkeys suffered any damage to their dopamine neurons. Next he tried varying the humidity, with equally negative results. Then he tested the hypothesis that male and female monkeys were affected differently, but this also turned out not to be the case.

It was beginning to look as if some fundamental error had been made in the published study. Had the drugs been wrongly prepared, such that the monkeys received more than the intended total dose of 6 milligrams per kilogram? To test this idea, Ricaurte undertook yet another set of experiments, in which the monkeys were given 12 milligrams per kilogram—double the previous dose. Yet even these monkeys suffered no damage to their dopamine systems. Finally, Ricaurte repeated the original published study with baboons, but these animals, just like the squirrel monkeys, experienced no dopamine injury.

Having exhausted other hypotheses, Ricaurte now began to suspect that there was something wrong with the MDMA—either the MDMA he had used in the original study, which had gravely impaired the monkeys' dopamine system, or the MDMA that he had used in his more recent experiments, which had not. Ricaurte had obtained all his MDMA from the Research Triangle Institute (RTI), a nonprofit corporation based in North Carolina that produces special-purpose drugs under contract with the U.S. government. Like most scientists, Ricaurte relied on his suppliers to verify the purity of their drugs and didn't check them himself.

The MDMA that Ricaurte used for his published study came from a 10-gram bottle that he received from RTI on April 27, 2000. That bottle had long since been used up and discarded, so its contents couldn't be tested. But RTI did still have some of the same batch in storage. That sample, as well as a sample of the batch that Ricaurte had used in the newer studies, tested correctly for authentic MDMA.

In perplexity, Ricaurte ordered authentic MDMA from yet another, newer batch, and tested it on a baboon: it too failed to damage the animal's dopamine neurons.

It then occurred to Ricaurte that, although the original bottle of MDMA had been discarded, he did still have the brains of two monkeys who had received injections from that bottle. These were the two monkeys (one squirrel monkey and one baboon) who had died shortly after receiving their third injections; Ricaurte had frozen their brains and put them in storage. Because these two animals had died so quickly, some of the MDMA should still have been present in the frozen brains. Ricaurte therefore tested for the presence of MDMA in these two brains as well as in the brains of other animals that he was certain had received authentic MDMA. He detected MDMA in those other animals readily enough, but there was no sign of MDMA in the two animals that had died after receiving the three injections from the suspect bottle. So Ricaurte concluded that the bottle, though labeled "MDMA," had not in fact contained that drug.

But *something* had to have been in the bottle, and that something had to have been a drug capable of severely damaging the monkeys' dopamine systems, as well as inflicting lesser damage on their serotonin systems. Ricaurte was very familiar with a likely candidate, if for no other reason than that Rick Doblin had forcefully reminded him of it in his letter to *Science* just a few weeks earlier. That drug was methamphetamine—speed. If Ricaurte had given his monkeys methamphetamine rather than MDMA, that would not only explain the damage to their dopamine system; it might also explain why two animals had died and two had fallen sick. That's because methamphetamine is a more potent drug than MDMA: When users take pure speed (often referred to as crystal meth or ice) they typically take no more than 100 milligrams, but Ricaurte had given his monkeys an amount that would correspond to a human dose of 150 milligrams, and he had given the animals this dose three times in a row in the course of just a few hours.

So Ricaurte wanted to test the frozen monkey brains for the presence of methamphetamine. But he wasn't very familiar with the testing protocol, so he started by taking some pure methamphetamine (from a bottle also supplied by RTI) and testing it. To his puzzlement, the test didn't seem to come out right—the results were more suggestive of MDMA than methamphetamine. Ricaurte therefore sent out a sample for testing by a much more sensitive procedure—mass spectrometry. The results were unambiguous: The substance was not methamphetamine but MDMA.

It was now beginning to seem that RTI had provided Ricaurte with incorrectly labeled drugs, and not just once but twice. And, even more remarkably, the two suspect bottles had been received from RTI on the same day, April 27, 2000, as part of the same order and in the same package. Was it possible that a technician at RTI had accidentally switched the labels between the two bottles so that the MDMA ended up being labeled as methamphetamine and vice versa?

By this time it was July of 2003, and Ricaurte's annual progress report to NIDA was due. But Ricaurte had no progress to report: He had shelved all the experiments that he had planned to do in that year in order to get to the root of MDMA mystery. So he wrote a report that described his failure to replicate his own study and the various attempts he had made to understand the reason.

Having sent off the report, Ricaurte did the clincher experiment: He took samples from the brains of the two monkeys who had died after their "MDMA" injections and sent them for analysis by mass spectrometry. The results were unambiguous: The brains contained methamphetamine, but no trace of MDMA. Clearly, the dopamine injury sustained by these and the other monkeys had been caused by an overdose of speed, not by Ecstasy. Ricaurte's 2002 *Science* paper was utterly and completely wrong, though apparently through no fault of his own.

At this point only a very few people—chiefly Ricaurte's research group, NIDA staff, and a few other colleagues—knew what had transpired. But Ricaurte now had to do what scientists dread ever having to do, which was to write a letter of retraction to his publisher, *Science*. The letter was not simply a retraction but a recounting of the entire investigation that had led him to conclude that RTI had provided him with mislabeled drugs. Interestingly, Ricaurte did not completely recant the conclusions of the retracted study; that is, he did not state that recreational doses of Ecstasy were harmless to dopamine neurons, even though his results indicated that they were. On the contrary, he cited other papers that suggested that Ecstasy was toxic to the dopamine system and hinted that future experiments would document the fact. It was an odd way to sign off on a letter of retraction—a kind of dogged "I'll be back and I'll prove I was right."

Science published Ricaurte's letter on September 12, 2003, but the news leaked out a few days earlier, and it caused a sensation, sparking articles in all the major national newspapers and also overseas. The *New York Times* didn't just run a news account of the event, it followed that up with an investigative article that alleged all kinds of other scientific misdeeds by Ricaurte. The *Times* cited experts who claimed that, at one time or another, Ricaurte had used inappropriate statistical procedures or "played games with his data." A drug user who had been a volunteer in one of Ricaurte's human studies described what seemed to be a variety of procedural lapses on Ricaurte's part, such as failing to test for undeclared current drug use. (This particular person had used heroin just five days before participating in the study, but his statement to the contrary was accepted at face value, he told the newspaper.) Richard Wurtman, director of clinical research at the Harvard/MIT health sciences division, told the *Times* that Ricaurte was "running a cottage industry showing that everything under the sun is neurotoxic."

Most of the scientists who expressed critical views to the *Times* and other media sources were those who, like Wurtman, had had longstanding disagreements with Ricaurte or who were hoping to use Ecstasy in clinical research. But even scientists who had initially praised Ricaurte's study expressed very different opinions when his retraction appeared. Colin Blakemore, a neuroscientist who was then at Oxford University but who now heads Britain's Medical Research Council, had expressed himself as follows to the *Telegraph* of London when Ricaurte's study was originally published: "This new study provides further evidence that Ecstasy can be toxic to nerve cells . . . I think people would be well advised to avoid it." And he cited unpublished work from his own laboratory as being consistent with Ricaurte's results. But after the retraction appeared, Blakemore commented (in an unpublished letter to the editor of *Science* that was quote by *The Scientist*) that "the study was so obviously flawed that even I (not a pharmacologist) picked up the problems as soon as I saw the paper." When I asked him in 2006 about the apparent inconsistency in his comments, Blakemore told me that he had made the earlier, favorable comments on the basis of an inaccurate press release and had not yet actually read Ricaurte's paper.

Ricaurte's retraction left a couple of loose ends untied. For one thing, Ricaurte mentioned in the retraction that one of his "MDMA"-treated animals—a baboon—received its drug from a different source than the allegedly mislabeled bottle. Presumably that animal received authentic MDMA and not methamphetamine, and it should therefore have been spared any damage to its dopamine system. Yet the published paper implied that *all* the animals were similarly affected. Grob picked up on this as another example of a lack of forthrightness in Ricaurte's work. "For many years he has had a pattern of being very selective as to the data he discloses and the data he elects not to disclose," Grob said.

When I asked Ricaurte himself about this animal, he responded a bit cryptically: "In that baboon the level of dopamine was—I don't recall the exact value, but it was not reduced to the extent that we had seen in the others, and it was difficult to discern from that single value whether it was in the control range or whether it was modestly reduced." In other words, that animal was uninformative: It was neither normal enough to raise a red flag about the genuineness of MDMA's apparent toxicity, nor abnormal enough to undermine the mislabeled-bottle hypothesis as an explanation for the erroneous findings.

Another and more remarkable loose end was this: Officials at RTI did not go along with Ricaurte's explanation for what had happened. After an internal investigation, their spokesman said that there was "no evidence" that the bottles had been mislabeled in the way that Ricaurte had deduced. Of course, there may be no evidence for any number of events that did actually happen. But when I asked an RTI spokesperson in 2006, he denied the company's responsibility more positively. "Although we do not know what might have happened to the materials after they were received by Ricaurte," he wrote, "we reject with certainty that we mislabeled [them]."

On the face of it, this leaves nothing but conspiracy-style explanations. "The only other way that something like that could have happened," Ricaurte told me, "would be if someone in my lab willingly went in and tried to take all of the contents of the bottle that was supposedly containing methamphetamine and pushed them into the bottle that contained MDMA. And I've asked my chemist friends how feasible would that be, without causing some cross-contamination, without disturbing the labeling of the bottles, and without exception chemists who know this business tell me that that would be nearly impossible to do."

So Ricaurte believes that the error did occur at RTI in spite of the company's denials. "You must recognize," he went on, "that RTI produces drugs not only for much of the animal research that goes on in the United States and around the world, but the drug supply for many human clinical studies, so in retrospect it was extremely naive on my part to begin to think that the drug supplier would acknowledge their—" Ricaurte broke off in mid-accusation and withdrew to safer ground. "But in some ways it didn't matter to me; the question was just identifying the error and making sure that colleagues in the scientific community [were informed], that the scientific record was corrected."

While the retraction of his 2002 paper must have caused Ricaurte a great deal of embarrassment and soul-searching, it does not seem to have affected his career in a major way. He remains an associate professor in good standing at Johns Hopkins University Medical School, and he still receives research funding from NIDA. In fact, his research continues very much in the same vein as before. The large number of experiments that he did in the quest to understand why he couldn't replicate the 2002 study were not wasted: They were mined for data that went into new papers. In 2005 Ricaurte and his colleagues published an expanded version of his 1998 study of the effects of Ecstasy on the serotonin system; he stuck to the same conclusions as before, although he was more open to idea that the damage reversed itself over time. And in the same year he published a study reporting that amphetamine—a legal prescription drug used in the treatment of attention deficit/hyperactivity disorder—caused damage to the dopaminergic system in monkeys that was similar in some respects to the damage caused by methamphetamine.

In spite of the continuation of Ricaurte's research, the retraction of the 2002 paper did seem to weaken the impact of his work in some respects. Most notably, researchers who wanted to test Ecstasy in the treatment of posttraumatic stress disorder, who had previously encountered roadblock after roadblock in the way of their efforts to begin their studies, suddenly found themselves in business. Doblin and Mithoefer's proposed study got Review Board approval just two weeks after Ricaurte's retraction was published. Doblin questions any causal connection between the two events, but Charles Grob told me that the approval was "clearly attributable to Ricaurte's work being seriously questioned." In any event, the study began in 2004 and, according to Doblin, by two years later it was showing a beneficial effect of the drug.

Ricaurte told me that he wasn't opposed to the Doblin/Mithoefer study so long as the subjects were properly informed of the risks. He added: "I would have thought that people like Rick Doblin would say, 'Gee, maybe we should take what's coming out of that laboratory more seriously—when they make mistakes, they acknowledge them.' But it's had completely the opposite effect. Doblin's a remarkable character in many ways. I think he truly thinks that he can make this a better world if everybody takes Ecstasy. I don't doubt that he's trying to be helpful. But if you're trying to do that, why would you not want people to be aware of any potential risk that your magic pill may have?" (I don't know any basis for the suggestion that Doblin does not want people to be aware of Ecstasy's risks.)

Ricaurte's retraction did not lead to any rethinking of the Illicit Drug Anti-Proliferation Act by Joe Biden or other Congressional leaders. The Act remains law, although it doesn't seem to have been enforced in any very energetic way. According to the DEA the use of Ecstasy by American youth declined significantly from 2002 onward, a change that the DEA attributes to public educational campaigns against the drug. These campaigns include the DEA's Web site, which still carries a description of Ricaurte's study, as presented by the DEA's director to Congress in 2002, without any mention of the fact that the study has been retracted.

Finally I asked Ricaurte, "What would you say to a teenager who said that he or she was thinking of trying Ecstasy?" I thought I was lobbing him a softball that he could swat out of the ballpark with a terse and quotable comment such as "Don't!" But he remained true to type. "You know it's really quite remarkable," he said. "It's remarkable to me how difficult it is to convey what in many ways I think is a very simple message that emerges from, gosh, almost two decades of research with MDMA and related amphetamine derivatives. What we know, or what I think we know, what I think we've learned over the last two decades is (a) MDMA has the potential to damage brain serotonin neurons in most every species that's been examined except the mouse, where it happens to damage dopamine systems, and we have to recognize that we don't know what the better animal model is—is it all of the others, just because we live in a democracy, or perhaps the mouse is more representative? I happen to think that the mouse is the outlier, but I don't know that for a fact, but I think we've learned that MDMA is a drug that has the potential to damage monoamine systems, serotonin systems, in most all the animal species tested, and I think the other thing we know, and I think the other thing that should be conveyed, is that you don't need heroic doses of the drug to produce this selective form of brain injury. What we've learned with MDMA is that the difference between the size of the toxic dose and the size of the pharmacologic or effective dose—that that difference seems to be very small. We don't know exactly what that margin of safety is, but we do know that it appears to be narrow. Where we do know that from? From a number of studies where people have now tested lower doses, single doses, giving the doses orally, in a way that the drug is used by humans, and collectively that data says, you don't need heroic doses, the margin of safety for this drug may be narrow. And it a nutshell I think those are the two things that people, that any people who are contemplating using MDMA ought to be aware of, just by way of making an informed decision about the drug they're about to take."

CHAPTER 5

ENGINEERING GEOLOGY: The Night the Dam Broke

The urban tentacles of Los Angeles have not yet reached San Francisquito Canyon, an arid corridor that slices southward through the Transverse Ranges, forty miles northwest of downtown. A hiker can walk for miles there and see no life beyond the circling hawks and an occasional rattlesnake. But he couldn't fail to notice the immense, eroded blocks of concrete, some weighing ten thousand tons or more, that lie half-hidden in the chaparral like the decaying ruins of a long-lost civilization. These monoliths serve as memorials to the victims of America's worst civil-engineering disaster of the twentieth century, the failure of the St. Francis Dam on March 12, 1928.

The dam was the brainchild of William Mulholland, the famed water engineer whose projects made possible the explosive growth of Los Angeles during the early part of the twentieth century. Born in Belfast in 1855, Mulholland arrived in Los Angeles at the age of twenty-two with little education and a few dollars in his pocket. He took employment as a ditch tender for the city's water system, which was then dependent on the erratic local supply provided by the Los Angeles River. He quickly demonstrated an extraordinary aptitude for water engineering, and he rose through the ranks of the water company to become its superintendent. In 1902, when the city of Los Angeles turned the company into a public utility, Mulholland was appointed its chief engineer and general manager.

Mulholland's most well-known achievement was the Los Angeles Aqueduct, begun in 1908, which brought water to the infant metropolis from the Owens River in the eastern Sierras. The aqueduct was an engineering marvel for its age, not only on account of its great length (233 miles), but also because, in spite of some very rugged terrain along the way, water moved from one end to the other entirely by the force of gravity. Where the aqueduct crossed a canyon, the water was carried down the slope, across the canyon floor, and up the other side within a U-shaped length of steel pipe called an inverted siphon; these siphons had to be tremendously strong to withstand the internal pressure generated by the drop. The aqueduct also had many tunnels: most notable among these was the five-mile-long Elizabeth Lake tunnel, near the head of the Antelope Valley. Within this tunnel the aqueduct passed directly through the San Andreas Fault, and the torturous geology at the Fault posed special engineering problems. After emerging from the tunnel the aqueduct headed southward along the eastern flank of San Francisquito Canyon toward Saugus, terminating in a set of small reservoirs in the San Fernando Valley and nearby areas.

Thirty thousand Angelenos turned out in November 1913 to celebrate the opening of the aqueduct. The benefits to the city quickly became apparent, as irrigated orchards sprang up in the San Fernando Valley and plentiful water became available for homes and gardens in new subdivisions. Even today the Los Angeles Aqueduct—extended in length and supplemented by an additional parallel pipe—supplies the bulk of the city's water. The system brings almost half a billion gallons of water from the eastern Sierras to the thirsty metropolis every day.

Bill Mulholland became an almost godlike figure to the citizens of Los Angeles. He was urged to run for mayor, but he turned down the suggestion with a memorable comment. "Gentlemen," he told a crowd, "I would rather give birth to a porcupine backwards than be mayor of Los Angeles." Instead, he devoted himself to improving on what he had achieved. Over the following decade he added four hydroelectric powerhouses to the aqueduct. Two of these were situated in San Francisquito Canyon: One was located at the head of the canyon, where the aqueduct dropped 1,000 feet after exiting the Elizabeth Lake tunnel, and the other was sited six miles down the canyon. Over time, the powerhouses generated enough electricity to defray the entire costs of the aqueduct's construction.

Godlike though Mulholland may have appeared to Angelenos, to the residents of Owens Valley he seemed more like the devil. The aqueduct sucked the lifeblood out of the valley. For every orchard that was planted in the San Fernando Valley, one had to be abandoned in Owens Valley. Agriculture came to a halt. Owens Lake, the natural terminus of the Owens River, dried up completely in 1924, and the lakebed gave off a cloud of arsenic-laced dust that choked the valley every time the wind blew. Meanwhile, the city's agents began purchasing water rights farther to the north, enriching some settlers but threatening the rest with penury.

Infuriated by these developments, some of the settlers began a campaign of sabotage. In 1924 they seized the aqueduct's headgates and diverted water back into the Owens River. This episode was followed by dynamite attacks on the inverted siphons and other parts of the aqueduct. The attacks led to interruptions in water delivery and necessitated expensive repairs.

The settlers' campaign failed. Although the city's acquisition of the local water rights involved some deception, it was done more or less in accordance with the legal requirements current at the time. The settlers gained little traction with the courts or with public opinion and eventually gave up. The environmental problems in Owens Valley have remained unresolved to the present day: The city of Los Angeles is making some attempts at remediation, such as covering parts of the bed of Owens Lake with gravel and diverting a small portion of the Aqueduct water back into the Owens River.

During the early 1920s, Mulholland came to realize that the city needed more water-storage capacity. One reason was the fear of drought. In fact, one three-year drought reduced flows to the point that the city had insufficient water to supply the needs of the farmers in the San Fernando Valley. In addition, there was the threat of sabotage. Finally, there was the always-looming danger of a rupture of the San Andreas Fault, an event that would block the aqueduct within the Elizabeth Lake tunnel. For all these reasons, Mulholland wanted to construct a set of reservoirs south of the fault that collectively could store a year's supply of water for the city. The largest of these reservoirs, designed to hold half of the entire supply, was to be sited in San Francisquito Canyon, and construction of the necessary dam—its name anglicized to "St. Francis"—began in 1924.

On the face of it, San Francisquito Canyon looked like an ideal location. In the northern part of the canyon lay a broad valley that could easily hold 32,000 acre feet (or about 10 billion gallons) of water. In fact, it had been the site of a large lake in prehistoric times. About halfway down the canyon, about a mile and a half north of the lower powerhouse, a rocky spur jutted out from the canyon's western flank, constricting the canyon to a gorge barely 200 feet wide at the canyon floor. Because the sides of the canyon were sloped, however, a dam would have to be considerably wider at its top—about 550 feet if the dam were built straight across, and more if it were curved. Thus the greater portion of the dam would consist of its abutments, the sections that rested on the canyon's sloping sides.

Mulholland had several geologists look at the proposed dam site before he made the decision to go ahead. These experts included John Branner, chairman of the geology department at Stanford University and the university's second president. The geologists gave Mulholland their opinion that the site was suitable, but their inspections may not have been very detailed. After the disaster, Mulholland cited these geologists' positive opinions but did not produce any written reports to back them up. Branner's approval probably took the form of a verbal "looks OK to me" after a single visit to the site. He died before Mulholland made his final decision on the dam's location.

One potentially troublesome feature of the site was already well known. A geological fault, the San Francisquito Fault, ran southward along the canyon, traversing the prospective site of the dam near the bottom of the sloping western wall of the gorge. Thus, if the dam were built, the fault would lie at the base of the dam's west abutment (or "right" abutment, according to the convention that the viewer is facing downstream). The fault was believed to be inactive, however, meaning that it was no longer subject to the slow accumulation of stress that might trigger an earthquake.

The reddish rock to the west of the fault, where the dam's right abutment was to be located, is a conglomerate—that is, it consists of pebbles and cobbles in a matrix of hardened sand or silt. This particular example is known as the Sespe Formation or Sespe Red Beds. The rock in this formation, especially in a zone near the fault, is rather crumbly and easily weathered, but its most startling behavior is evident when it becomes wet. When I visited the dam site in 2006, I pulled a loaf-size chunk of rock out of the slope where the right abutment once stood and placed it in the creek. The rock underwent a kind of slow-motion explosion: Over a period of about ten minutes it gave off bubbles, started cracking, and then fell apart into a heap of stones and mud. In spite of this behavior, Mulholland conducted tests that convinced him that the foundations of the right abutment would not fail or allow water to percolate through.

The rock to the east of the fault, which would carry the central section of the dam and its left abutment, is a mica schist. This is a sedimentary rock that has been altered by heat, pressure, and shearing forces so that its constituent grains are flattened, which gives the rock a laminated structure rather like slate. This particular formation is named the Pelona Schist. It is much harder than the rock of the Sespe Formation, but its layered structure gives it the tendency to split and slide along the plane of the layers, like a deck of cards. What was worse, the rock layers were tilted such that they roughly paralleled the slope of the east side of the gorge. At the fault itself, where the Sespe Formation and the Pelona Schist met, there was a layer of claylike "fault gouge," about eight inches wide, that had been generated during innumerable ancient ruptures of the fault.

Undaunted by the problematic geological features of the site, Mulholland decided to go ahead. As the "chief," his word was as good as law, and any review of his decision within or outside of the water department was perfunctory if it happened at all. As for environmental reviews or state safety inspections, such things did not happen in the 1920s.

Mulholland constructed the dam from concrete. Most of his prior dam-building experience involved earthen dams, but one year before starting the St. Francis Dam he had begun work on his first concrete dam. This was the dam now known as the Mulholland Dam, which confines the relatively small Hollywood Reservoir on the western fringe of Griffith Park. Apparently Mulholland chose concrete because of a lack of clay—needed to form the water-resistant core of an earthen dam—at the Hollywood and St. Francis sites. The design for the St. Francis Dam was very similar to that for the Mulholland Dam, and both were based on the then-current textbooks of dam design.

The two dams were gravity-arch dams. This means that they depended primarily on their weight to hold back the water in the reservoir. This weight—a quarter of a million metric tons in the case of the St. Francis Dam—thrusts directly downward, whereas the hydrostatic pressure of the reservoir water thrusts more or less horizontally downstream. The resulting combined thrust is aimed obliquely downward. For the dam to be stable against tilting, the combined thrust must be directed into the bedrock within the middle third of the dam's footprint, not near the downstream edge or "toe" of the dam or, even worse, downstream of the toe. The plans developed by Mulholland's design engineers met this criterion, of course. An additional safety factor was added by the dams' arched shape (curved convexly into the reservoir). This had the effect that some portion of the reservoir's horizontal thrust was carried sideways into the dam's abutments and thus into the canyon walls.

Conservative though this design was, Mulholland made it much less so by changes that he ordered after construction got under way. For one thing, he twice raised the height of the dam, from the original 175 feet to 185 feet and then to 195 feet—an 11-percent increase in height. His aim, of course, was to increase the holding capacity of the reservoir. But Mulholland did not order any compensatory thickening of the dam at its base. In fact, he actually omitted the lowermost portion of the dam's toe, leaving the dam 20 feet (or 11 percent) thinner at its base than called for in the design. This latter fact was only discovered decades later by Charles Outland, a local historian who made an extensive study of the disaster and wrote a detailed book about it. Outland spotted the shortened toe by closely examining photographs that had been taken during the dam's construction. Because the dam was

taller, and thinner at its base, than the original design had specified, it was significantly less stable against tilting.

The dam was completed in May 1926. Filling of the reservoir
began during the construction phase but was not complete until
March 7, 1928. At that point, the water lapped just a few inches
below the spillway. Photographs taken at that time convey an image
of graceful strength. The dam's downstream face, rather than being
smooth like most high dams that we're familiar with today, was
stepped. This gave the dam the look of a Roman amphitheater and
emphasized its curvature. Behind the dam, the reservoir spread out
into the broad upper San Francisquito Canyon and its side-bays: It
resembled a natural and serene lake.

Day-to-day inspections of the dam were left in the hands of the
damkeeper, Tony Harnischfeger, who lived with his girlfriend, Leona
Johnson, in a cottage located in the canyon a few hundred yards
downstream from the dam. On the morning of March 12, 1928, five
days after filling of the dam was complete, Harnischfeger telephoned
Mulholland to tell him that muddy water was leaking from the base
of the dam's right abutment. High dams usually leak a certain amount
of water, and the St. Francis Dam had already sprung several small
leaks during the filling process, but the fact that this leak was muddy
was novel and ominous. It suggested that water was not merely passing through the dam but was removing material as it did so.

Mulholland and his chief assistant, Harvey Van Norman, immediately drove from his downtown office to the dam site, which they
reached around ten thirty a.m. They confirmed that muddy water
was indeed flowing down the foundations of the right abutment, at
a rate of about fifteen gallons per second. When they clambered up
to the site where the water was emerging from the Sespe Formation,
however, they saw that the water was clear. The mud was mixing in
as the water ran down the slope and across the embankment of a
dirt road. Some water was also leaking from the base of the other,
left, abutment, but this water was also clear. Relieved to find that the
dam was not being undermined, Mulholland made a mental note to
have the leaks repaired at some later date, and he and Van Norman
returned to the city.

The dam broke at two and a half minutes before midnight that evening. There were no surviving eyewitnesses to the collapse, but the exact time could be pinned down because it caused an interruption in the transmission of electric power in lines that ran down the canyon. The interruption was experienced in Los Angeles only as a two-second dimming of lights, but areas closer to the dam were plunged into darkness.

It is likely that Harnischfeger and Johnson did witness the dam's collapse before they became its first two victims. Leona Johnson's fully clothed body was later found trapped amongst the slabs of broken concrete at the base of the dam. Most probably Harnischfeger and Johnson saw or heard some premonitory sign of the collapse, dressed, and walked up to the dam to see what was wrong, only to be caught by a cascading mass of water and concrete. Harnischfeger's body was never found, nor was that of his young son.

A few survivors did hear or feel the collapse. A motorcyclist named Ace Hopewell was driving up the valley shortly before midnight. The road ran along the canyon's east wall: At the point that it passed the dam, it was cut into the Pelona Schist only thirteen feet above the dam's left abutment. Hopewell noticed nothing amiss as he passed the dam, but a mile farther up the road he heard a sound as of rocks falling. He stopped for fear of running into a landslide, but the sound seemed to be coming from behind him, so he continued on his way.

E. H. Thomas was one of the staff of the lower powerhouse, which was situated in the canyon a mile and a half below the dam. His particular job, however, was to attend to the surge tank, high on the east rim of the canyon. The surge tank's function was to damp out violent changes in hydrostatic pressure caused by the opening and closing of valves; from the surge tank, pipes (or "penstocks") carried the aqueduct water down to the twin generators in the main powerhouse building. Thomas lived with his mother in a cottage not far from the surge tank. At about midnight the two were woken by a strong shock, followed by a continuous vibration. They first assumed that it was an earthquake. Thomas dressed, took a flashlight, and made his way toward the tops of the penstocks. Looking down into the canyon, he saw nothing but rushing water. The powerhouse, if it still existed, was completely overtopped by the flood, which scoured the canyon walls to a height of 120 feet above the creek. Thomas realized that the twenty-eight other powerhouse workers and their families, whose homes were in the floor of the canyon, had most likely been swept away to their deaths.

Actually, three people did survive. One was Ray Rising. "I heard a roaring like a cyclone," he said later. "The water was so high, we couldn't get out the front door. . . . In the darkness I became tangled in an oak tree, fought clear, and swam to the surface. . . . I grabbed the roof of another house, jumping off when it floated to the hillside. . . . There was no moon and it was overcast with an eerie fog—very cold." Rising met up with a worker's wife, Lillian Curtis, and her young son, who had had similar narrow escapes, and the three of them waited for the dawn together, but his own wife and three daughters died, along with sixty-four other members of the powerhouse community. The powerhouse itself, a 65-foot-high concrete building, was swept away—only the generators remained in place, half-buried in mud.

The floodwaters rushed onward. About fifteen minutes after the dam broke, the flood reached a cattle ranch near the southern end of the canyon. All the buildings were swept away, along with several of the residents, but a few had been woken by the noise and were able to scramble to higher ground.

At the southern end of San Francisquito Canyon, the creek joins with those from several other canyons to form the Santa Clara River, which runs forty miles westward to the Pacific Ocean south of Ventura. Near where the creeks meet lay the small community of Castaic Junction—basically an auto park, with tourist cabins, a gas station, a café, and a few other buildings. It was located just north of the present-day site of the Magic Mountain theme park, and it served travelers on the highway that connected Los Angeles with California's Central Valley—the predecessor of today's Interstate 5.

It took fifty minutes for the floodwaters to reach Castaic Junction, but they had lost little of their ferocity along the way. George

McIntyre, the nineteen-year-old son of the auto park's owner, was alerted by the rushing, crashing noises coming from the east. He, his father, and their cook went outside to see what was happening. They saw bright flashes from the direction of the Edison station at Saugus, and they concluded that there was something amiss there. Then they watched with amazement as one of the tourist cabins began moving off its foundations. Within moments the father and son were knocked off their feet by the floodwaters. While holding on to each other, the two men were swept away, but not before they caught a final glimpse of George's younger brother struggling to get through the window of one of the cabins. After a while George and his father were able to grasp onto a utility pole, but the father was injured, probably by floating debris, and after muttering a brief good-bye to his son he let go his grip. That was the last George saw of him. Eventually George also let go of the pole and was sucked deep into muddy water. At long last he found himself back at the surface, half-choked with water and mud. After drifting for some time he collided with the branches of a cottonwood tree, where he was able to hold on until the flood had receded. Besides George McIntyre, only one other person survived from the Castaic Junction community—the cook, who escaped the floodwaters by running to higher ground. All the structures were completely destroyed, and four miles of the north–south highway were inundated.

By this time, the wider world was beginning to find out about the catastrophe. The initial break on the Edison line had been followed by a cascade of wider electrical failures, as more lines were brought down, switching stations were shorted out, and emergency connections were overloaded. Soon the city of Los Angeles, the Antelope Valley, and all the coastal cities north to Santa Barbara had lost power. Charles Heath, Edison's superintendent of transmission, later testified that he guessed already at 12:05 a.m. that the St. Francis Dam had failed, and that, after telling his subordinates to warn the towns in the Santa Clara Valley, he set out for Saugus in his official car, with siren blaring and red light flashing. He said that he reached Saugus at about 12:45—just about the same time that the floodwaters were entering the head of the valley.

Heath knew that one hundred fifty Edison workers were sleeping in tents at a construction camp on the bank of the Santa Clara River, eight miles down the valley. He couldn't reach them, because the road and the bridges were out, so he attempted to telephone the camp from the Saugus substation, which itself was being inundated by the rising water. No one answered the phone; then, after several repeated attempts, the line went dead. The flood, forming a wall of water forty feet high or so, had struck the camp without any warning. Of the hundred and fifty workers sleeping there, eighty-five died. Few of them had even been able to get out of their tents.

Further down the valley were the towns of Fillmore and Santa Paula. The destruction of roads, bridges, power lines, and telephone lines had cut them off from any communication to the east, but at 1:30 a.m. the night telephone operator at Santa Paula received a call from the coast: The St. Francis Dam, she was told, had broken and floodwaters were descending the Santa Clara Valley. Working by candlelight, the operator began calling local police officers, town officials, and the residents of the town whose homes were closest to the riverbed. Two police officers drove through the streets on their motorcycles with sirens blaring. They stopped at every third house or so, warning the sleepy occupants of the oncoming flood and telling them to alert their neighbors and move to higher ground. Soon the entire town was on the move. Meanwhile, squad cars raced up the valley to Fillmore and alerted the population there. As the cars attempted to drive farther east, however, they were stopped by the arriving flood, which filled the entire two-mile width of the valley and was advancing westward at about 12 mph.

Much of the Santa Clara Valley was used for citrus farming and other tree crops. The groves were obliterated by the oncoming water and mud. In addition, the low-lying portions of the valley towns, especially Santa Paula, were inundated. Houses were carried off, to be smashed up by the roiling water or to be left reasonably intact but several blocks away from their foundations. The bridges across the river were destroyed: The bridge at Santa Paula was demolished just minutes after a police officer ordered a crowd of would-be spectators to get off it.

As the floodwaters moved westward, they also slowed and spread out. By the time they reached the coast, five hours after the dam broke, they were moving at no more than a jogging pace. The coastal city of Oxnard was evacuated, but that turned out to be unnecessary.

The total death toll was estimated to be between 400 and 450 persons. This figure may be an underestimate, given that then, as now, many undocumented migrants from south of the border lived in the low-lying areas of the canyons and riverbeds. Though impromptu morgues overflowed with bodies, many of the dead were never found. Some were undoubtedly swept out to sea: Bodies washed up on beaches as far south as San Diego. Others were buried under many feet of mud. In addition to the human toll, there was enormous destruction of livestock, buildings, infrastructure, orchards, and farmland.

Eventually, the city of Los Angeles settled with most of the victims of the disaster without litigation. Persons who had lost family members typically received $10,000 to $20,000—or $100,000 to $200,000 in today's dollars. In one case that did go to court, Ray Rising, the sole worker to survive at the lower powerhouse, was awarded $30,000 for the loss of his wife and three daughters.

What caused the dam to fail? William Mulholland himself, accompanied as always by Harvey Van Norman, rushed to the dam site in the small hours of the morning after the disaster. As dawn broke, an extraordinary scene revealed itself. A slim central section of the dam still stood in its original place, ranging 200 feet above the empty reservoir like a lone incisor in an otherwise toothless jaw. To its left, another huge section of the dam had slumped downward and across the surviving upright section, shearing off much of its stepped downstream face. This section had broken into three or four giant blocks. But the remainder of the dam, including its left and right abutments, were simply gone—carried as much as a mile down the canyon by the floodwaters. The rocky foundations of the abutments had also been scoured away to a depth of twenty feet or more. The left wall of the canyon had fallen away in a giant landslide, carrying a stretch of the roadway with it. Evidently the material in this slide had mixed in with the floodwaters, turning them into a slurry that was dense enough to freight the huge blocks of concrete far down the canyon.

Surveying the scene, Mulholland immediately suspected sabotage. After all, the aqueduct itself had been the object of dynamite attacks just a few months earlier, so why not the dam, too? Mulholland mentioned or hinted at sabotage as the cause on several occasions, such as at the coroner's inquest on the victims of the disaster. Some experts presented what purported to be evidence of sabotage, such as the presence of dead fish that were supposedly stunned by an explosion.

Still, the sabotage theory never really took hold, and with good reason. It would have been an enormous undertaking to bring down the dam, far beyond what could have been accomplished unnoticed by a small team of saboteurs. Furthermore, the seismological station at Caltech would have recorded the tremors induced by the blast, but inspection of the records showed nothing unusual at the time of the failure. To his credit, Mulholland did not use the sabotage theory to weasel out of responsibility for what had happened. He told the investigators, "If there is an error of human judgment, I am the human," and he was so laden by guilt that he professed an envy for those who had died.

There were, of course, investigations into the cause of the dam's failure. The most high-level of these was a commission of inquiry set up by the governor of California. The commissioners consisted of six engineers and geologists, including professors from Caltech and the University of California, Berkeley. Perhaps responding to the pressure of a public outcry, the commissioners carried out their task with extraordinary speed. They met for the first time on March 19, made a single visit to the dam site, and issued their final report on March 24, just twelve days after the disaster.

During their visit to the dam site, the commissioners noted the unusual characteristics of the Sespe conglomerate near the San Francisquito Fault. "When entirely dry it is hard and rock-like in appearance," they wrote, but in reality it was "held together merely by films of clay," and when placed in water it "quickly softened and turned into a mushy or granular mass." The commissioners also had compression tests done, which revealed that, even when dry, the rock of the Sespe conglomerate in the dam's right foundation was far weaker than the concrete of the dam itself.

It so happened that the remaining intact segment of the dam carried a gauge whose function had been to measure the depth of water in the reservoir. The depth was continuously recorded on a drum that was rotated by a clockwork mechanism. When the commissioners inspected the drum, they found that reservoir level appeared to have begun falling at about eleven thirty p.m. on the evening of the disaster and had fallen by about four inches by the time of the collapse. This suggested that there was an initial small breach in the dam or its foundations that had gradually widened itself until the overlying section of the dam lost its support and suddenly collapsed into the breach.

According to the commissioners' final report, the disaster was not caused by any shortcomings in the design of the dam itself, but by defects in its foundations—most particularly, by the weakness of the Sespe conglomerate, especially when wet. Because some of the segments from the right (western) side of the dam had been swept a long way downstream, the commissioners concluded that the initial break had occurred on that side, probably at or near the location where the San Francisquito Fault passed under the dam. Reservoir water had eaten away at the fault gouge, or the weak conglomerate near it, until the dam's right abutment had collapsed. The torrent of water, curving around the center of the dam in a giant eddy, had eroded the foundations of the dam's left abutment until it, too, collapsed, which in turn provoked the landslide on the canyon's western slope.

Given the nature of the rock of the Sespe Formation, the commissioners concluded that failure of the dam was inevitable, unless water could have been prevented from reaching the dam's foundation. They mentioned several steps that could have been taken to slow down the entry of water. These included the construction of a deep cut-off wall (a concrete wall built deep into the foundations at the upstream face of the dam), pressure grouting of the foundations (to fill spaces that would have permitted the entry of water), the drilling of drainage wells to remove water that did enter the foundations, and the construction of inspection galleries within the dam that would have allowed the state of its foundations to be monitored. But these steps, they believed, would only have postponed the dam's ultimate collapse.

In spite of Mulholland's suspicions about sabotage, he evidently did accept the notion that inherent problems with the dam or its foundations were the more likely cause of the disaster. That raised an urgent problem, because the near-twin of the St. Francis Dam was holding back the now-full Hollywood Reservoir. And the reservoir lay, not in some undeveloped canyon forty miles from Los Angeles, but immediately above residential neighborhoods within the city itself. To ward off a repeat of the disaster, Mulholland ordered a lowering of the reservoir.

The theory put forward by the commission—that the weakness of the Sespe conglomerate and its poor response to wetting were the prime reasons for the dam's failure—was widely accepted. It was used, for example, at the coroner's inquest on the victims, as part of an (unsuccessful) attempt to bring criminal charges against Mulholland. Even today there are experts who agree pretty much word for word with the commission's report. Geologist Jack Green of California State University at Long Beach, for example, has a Web page devoted to the disaster in which he states that the dam broke because water eroded through the dam's foundations in the neighborhood of the fault, just as the commission concluded.

Even in 1928, however, there were experts who voiced disagreement with the commission's theory of the disaster. Carl Grunsky was a well-respected consulting engineer based in San Francisco who had been retained by local ranchers during the construction of the dam, and who therefore had good opportunity to study it. Two months after the disaster Grunsky, along with his son and a Stanford geologist, presented an account of what had happened that differed markedly from that of the official commission. They argued that the dam broke because it was subjected to a kind of pincer action. As the dam was filled, the Sespe conglomerate began to swell as water invaded it, so the right abutment attempted to push the dam leftward. Meanwhile, the foundations of the left abutment responded to the pressure of the dam by starting to slide. This incipient landslide was slight enough to be unnoticeable, but it pushed the dam rightward. These forces, building gradually over the months prior to the failure, caused cracks to develop that partially separated the center of the dam from its two abutments. These cracks had been seen, but they were attributed to irregular contraction of the concrete during drying, and they did not cause any concern.

The drop in the level of the reservoir during the thirty to forty minutes prior to the dam's collapse, as recorded by the gauge, was an illusion, according to Grunsky. He calculated that for the drop to be real, such an enormous amount of water would have been flowing past the lower powerhouse that it could not have escaped notice, and the workers and their families would have had plenty of time to save themselves. In fact, the operator of the upper powerhouse, a man named Henry Silvey, had spoken by telephone with the duty officer at the lower powerhouse just ten minutes before the dam collapsed, and he did not report anything amiss.

What, then, had caused the appearance of a drop in water level? Grunsky found what he considered a decisive clue: While inspecting the remaining standing block of the dam, he noticed a ladder that had become trapped and crushed within a horizontal crack in its upstream face. The crack must have opened by several inches to allow the ladder to enter; in other words, the dam must have tilted upward and forward. According to Grunsky it was this upward tilting of the dam, caused by the pincer action described above, that had raised the gauge slightly out of the water, producing an illusory record of a loss of water prior to the collapse.

A further clue relating to the possibility of an incipient landslide was reported by Charles Outland, the historian mentioned earlier. Thirty-four years after the dam collapsed, a member of the staff of the upper powerhouse (probably Henry Silvey, but he requested anonymity) told Outland that he had driven up the canyon on the evening of the disaster. As he passed the left abutment of the dam, he was disturbed to notice that the roadway had sagged downward by up to twelve inches, producing a scarp of that height that he had to negotiate with great care. Given that the road was cut into bedrock

just above the dam's left abutment, the presence of the scarp suggested that a large block of the canyon slope under the dam's left
abutment had begun a downward slide well before there was any
other sign of the dam's failure.

Around 1990, a new analysis of the disaster was undertaken by J.
David Rogers, then an engineering geologist in private practice in the
Bay Area and Los Angeles. (Rogers is now an associate professor at
the University of Missouri, Rolla.) First, Rogers pointed out that the
east wall of the canyon, at the point where the dam was built, had
been the site of an enormous natural landslide in prehistoric times,
in which a giant sector of the mica schist had slid and tilted down,
ending up against the Sespe conglomerate and completely blocking
the canyon. It was this episode that created the ancient lake that once
filled the upper canyon. Then, at some later time, the rising waters
breached this natural dam: The lake emptied, leaving the broad,
sediment-filled valley that struck Mulholland as so suitable for a
reservoir. Thus, the dam's left abutment was built on rock that was
inherently unstable and liable to experience another slide, especially
when subjected to the pressure of the dam and the lubricating effect
of water that entered the schist. Rogers agreed with Grunsky that an
incipient slide in the rocky foundations of the left abutment was a
key causative factor in the dam's collapse.

Another factor on which Rogers placed a great deal of emphasis was what is called "hydrostatic uplift." This has to do with that
staple of elementary-school physics, Archimedes' principle. The principle can be stated thus: a body immersed in a fluid experiences a
buoyant force equal to the weight of the fluid displaced. This is why
bodies that are less dense than water float at the water's surface. But
even bodies that are denser than water, such as those made of concrete, will experience a reduction in their apparent weight when they
are immersed in water.

As long as a gravity dam rests on dry foundations, the full weight
of the dam thrusts downward. If reservoir water enters the dam or its
foundations, however, the dam is effectively immersed in the water.
The dam experiences a buoyant force equal to the hydrostatic pressure of the reservoir's water column, and this upward force nullifies

a significant portion of the dam's weight. According to Rogers's calculations, if the St. Francis Dam experienced full hydrostatic uplift, its apparent weight would have decreased by about 45 percent. As a result, the thrust resulting from the combination of the dam's weight and the reservoir's horizontal pressure would no longer have been directed into the bedrock within the middle third of the dam's front-to-back extent, as called for in the dam's design. Rather, it would be shifted 240 feet downstream—well beyond the toe of the dam.

The theory of hydrostatic uplift, as applied to dam construction, was poorly understood in the 1920s—the textbooks on which Mulholland's design engineers relied barely mentioned the topic. That is probably why Mulholland did not take effective precautions to prevent reservoir water from entering the dam's foundations: He was concerned about such percolation only insofar as it might erode the foundations, not as a factor tending to lift the dam.

Mulholland did install a few relief wells in the dam to drain off any water that seeped under it. These wells were placed only under the relatively narrow central section of the dam, however—the portion that rested on the canyon floor. There were no relief wells in the dam's abutments, which together formed the bulk of the dam's width. Yet abutments are readily affected by uplift, because as they climb the sloping sides of the canyon, their height and weight decrease so that the stabilizing downward thrust due to gravity becomes less.

Rogers found a photograph of the dam, taken some months before the filling of the reservoir was complete, in which the line of contact between the left abutment and the Pelona Schist, on the downstream face of the dam, was clearly wet. This observation, combined with the leakage of water in that area noted by Mulholland on the morning before the failure, is evidence that reservoir water had not merely entered the mica schist but had percolated all the way through to the downstream face of the left abutment. This water was not only facilitating slippage of the ancient landslide but also was exerting a gradually increasing uplift on the abutment. This tended to separate the abutment from the more stable central section of the dam.

As the mica schist crept progressively downward during the days and hours before the collapse, the incipient landslide pressed against the dam. Because of the dam's curvature, the abutments were oriented quite obliquely to the canyon walls. The pressure of the slide was therefore exerted primarily on the upstream face of the left abutment, putting the entire upstream face of the dam into tension. A short while before the final collapse, in Rogers's interpretation, a fairly small block of concrete near the base of the left abutment fell out of the dam. (This block was later found farther downstream than any other block, suggesting that it was the first to yield.) Water poured through this orifice: It undercut the mica schist, and it also entered transverse cracks in the dam, causing the central portion of the dam to experience full hydrostatic uplift and therefore to tilt upward by a few inches. This was what caused the water gauge to record an apparent fall in the reservoir level.

Then, at 11:57.30 p.m.—the time defined by the power outage—the schist was undercut to the point that a giant landslide occurred, involving about 900,000 tons of rock. The landslide destroyed the remaining part of the left abutment, but the material in the slide plugged the gap in the dam for a short time. Then water tore through the slide and the catastrophic emptying of the reservoir began.

Rogers pointed to several observations that supported the idea of a landslide-induced breach in the left abutment as being the first event in the dam's collapse, rather than a failure of the Sespe conglomerate under the *right* abutment as the governor's commission had concluded. For one thing, after the disaster a line of debris was found on the western shore of the reservoir near the dam, and this debris extended for several feet above what had been the reservoir's highwater level. Rogers concluded that this line of debris was thrown up by a large wave—a tsunami, essentially—generated by the landslide. For the debris to have been left above the high-water level, the reservoir must have been full, or very nearly so, at the time the landslide occurred. Furthermore, although both abutments eventually failed, the rocky foundations under the right abutment were scoured away much less than under the left foundations, even though the Sespe conglomerate was relatively weak. This indicated that the reservoir level was already relatively low when the right abutment breached. As a further sign that the right abutment failed late in the proceedings, the scour level along the west side of the canyon just downstream of the dam was much lower than it should have been if water from a full reservoir had been pouring through the right abutment. A final clue was offered by a thirty-foot-long pipe, which was attached to the underside of the water gauge and which ran down the upstream face of the dam. After the collapse, the pipe was visibly bent in the direction of the left abutment. This happened, according to Rogers, because while the reservoir level was nearly full, water was rushing toward the breach in the left abutment caused by the landslide.

The complete order of events, as visualized by Rogers, was roughly as follows. Reservoir water entered the foundations of the left abutment, causing hydrostatic uplift of the abutment and promoting an incipient landslide. Failure of a segment of the dam near the base of the abutment opened an orifice that allowed high-pressure water to extend the hydrostatic uplift to the central section of the dam and to trigger a massive landslide that collapsed the entire abutment. As the landslide material was washed away, a scouring of the dam's foundations caused the leftmost part of the central sector of the dam to collapse. The remaining portion of the center of the dam did not fall, but it tilted and (as was determined by triangulation after the disaster) moved slightly to the east. This caused a separation from the right abutment, which therefore lost its stability and collapsed, allowing water to pour through on that side too. Rogers calculated that the maximum rate of flow past the collapsed dam and down the canyon was about 1.7 million cubic feet per second, which is nearly three times the average flow of the Mississippi River at New Orleans. The reservoir emptied in less than an hour.

The Los Angeles Department of Water and Power soon replaced the broken dam with a new, earthen dam in a nearby canyon, and with time the St. Francis Dam and its tragic demise faded from memory. Still, the disaster had wide-ranging consequence for dam-building elsewhere. For a start, several committees looked into the question of what to do with the Mulholland Dam. Eventually, it was decided that the dam could be operated at a lower reservoir level, but as a precaution against failure 300,000 cubic yards of earth fill were placed against the dam's downstream face, completely burying its elegantly stepped facade. The construction of a much higher dam in San Gabriel Canyon, east of Los Angeles, was halted when a Berkeley geologist discovered that the western wall of the canyon, much like San Francisquito Canyon, was the site of an ancient landslide.

The disaster also caused great concern for the designers of the Hoover (or Boulder) Dam on the Colorado River, then in the planning stage. Politicians opposed to the dam, such as the governor of Arizona, used the disaster in their campaign to prevent the dam's construction. Although the dam, when finally built, incorporated some design changes that took account of the failure of the St. Francis Dam, it nevertheless experienced some percolation of reservoir water into its foundations, and gradually increasing hydrostatic uplift pressures were measured. Finally, in the 1950s, a program of pressure grouting of the dam's foundations reduced seepage to an acceptable level.

A tragic repeat of the St. Francis disaster occurred in 1959. The Malpasset Dam, near Fréjus in the south of France, collapsed when the reservoir was filled for the first time, killing between 400 and 500 people. As with the St. Francis Dam, it appears that the collapse was caused by hydrostatic uplift of the dam's left abutment, according to an analysis by Electricité de France.

Today, the problem of hydrostatic uplift is well understood, and extensive steps are taken during a dam's design and construction to prevent seepage of water under a dam, to drain whatever water does penetrate, and to monitor uplift pressures. Still, other modes of failure are possible. If water enters a dam's reservoir faster than the sluicegates or spillway can discharge it, for example, the reservoir will overflow the dam and likely destroy it. This occurred in China's Henan Province in August 1985. Storms that had been spun off by a typhoon dropped forty inches of rain on the area within the span of three days. A total of sixty-two different dams on two rivers were overflowed and collapsed in a chain-reaction disaster that cost the lives of an estimated 85,000 people.

After the failure of the St. Francis Dam and the subsequent inquiries, William Mulholland resigned his position as chief engineer and general manager of the Department of Water and Power. Already in his seventies and beset by a neurological condition that may have been Parkinson's disease, Mulholland lived the remaining seven years of his life out of the public eye. He is often described as a "broken man" in his final years. Considering the torrent of verbal abuse that Mulholland experienced after the disaster, it would not be surprising if his spirit had been broken, yet it was not, according to a memoir penned by his granddaughter Catherine Mulholland. Catherine describes her conversations with William Mulholland's nephew, also named William, who worked with him and knew him intimately as a family member. "He was not broken by that mishap," the nephew told Catherine, "because he never accepted the responsibility of something that was beyond his power."

CHAPTER 6

GENE THERAPY:
The Genes of Death

Before there were stem cells, there was gene therapy. The field took off in 1990, when geneticist William French Anderson of the University of Southern California reported that he had cured a four-year-old girl of "bubble-boy disease"—severe combined immunodeficiency, or SCID—by transferring the missing gene into her body. Soon, the idea of giving people new genes became the white-hot frontier of medical research. Touted as a possible cure for cancer, heart disease, diabetes, and hundreds of other conditions, this form of treatment was on everyone's lips, and nowhere more so than at the University of Pennsylvania's Institute for Human Gene Therapy, which was founded in 1993.

The Institute's director, physician and molecular geneticist James Wilson, led a team that had developed a potential treatment for an inherited disorder called ornithine transcarbamylase (OTC) deficiency. In baby boys who are born with this condition, their livers cannot metabolize the ammonia that they naturally produce when they digest protein, so ammonia levels in the babies' blood rise as soon as they have their first meal. Because ammonia is highly toxic to the brain, they quickly go into a coma and die. Wilson and his colleagues had engineered an adenovirus—a kind of common cold virus—to carry a normal version of the gene that is defective in the affected babies. The idea was to infect the babies with this modified virus (or "vector"), with the hope that some of the children's liver cells would take up the artificial gene and use it, at least temporarily, to replace the function of the defective one.

As with any new treatment, this one involved some risk to the subjects who participated in the initial clinical trials. Thus the question arose as to whether it would be ethically appropriate to test the new treatment on OTC-deficient babies. Wilson discussed this issue with Arthur Caplan, a bioethics specialist who was then on the staff of Wilson's institute. (He now heads the university's Center for Bioethics.)

In a fateful turn, Caplan advised Wilson not to test the treatment on babies, but on adults who had a less severe form of the disease. According to a 1999 article in the *New York Times,* Caplan gave that advice because he thought that the parents of extremely sick infants could not give informed consent: "They are coerced by the disease of the child," he told the newspaper. When I talked with Caplan in 2006, however, he denied that this had ever been his reason; instead, he said it was a simple matter of the federal regulations that were then in force. In an initial, or "phase-1," clinical trial the focus is entirely on testing for safety, and there is therefore no prospect of benefit to the subject, he said. In those circumstances regulations don't allow for the use of babies as subjects if there is any possibility of using adults.

Caplan was not entirely right about this. Although safety is indeed supposed to be the focus of a phase-1 trial, the Penn researchers did envisage that OTC-deficient babies might benefit from participation. One of Wilson's collaborators later told *Science* that the hope had been that the adenovirus infusion would bring the babies out of coma and keep them in reasonable health for a period of weeks or months, during which time other therapies might be brought to bear that would stabilize the children for the longer term. If that was so, the balance did not swing so decisively toward using adults in the trial.

Caplan offered another justification for his opinion, however. He said that it would have been impractical to do a clinical trial with OTC-deficient babies because of the emergency situation that arises when they are born. "What you'd have to do is fly in, enroll someone in a phase-1 trial within an hour—because you don't have a lot of time here, and you're going to show up out of the blue when they're expecting a healthy kid—and say, 'We just flew in, here's the liver surgeon, your baby's going to die, would you like to be in an experiment where there's going to be no benefit?'"

If OTC deficiency kills baby boys at the very dawn of their lives, who were the OTC-deficient adults who would be available for recruitment into the study? For the most part, they were women. The OTC gene is located on the X chromosome, of which males possess one copy and females two. Females who have a mutation in the OTC gene on one of their X chromosomes usually have a normal copy of the gene on the other chromosome, and this normal gene offers them partial or complete protection. (This situation is similar to that of other X-linked disorders such as hemophilia.) Female children may have no symptoms at all, or they may have mild symptoms that can be controlled by diet and medication. There are also rare instances of males whose tissues are a genetic mix or "mosaic" of normal cells and cells that are OTC-deficient. Again, such males tend to have mild symptoms that allow them to survive with proper medical care.

Enter Jesse Gelsinger. Jesse was born in June of 1981, the son of Paul Gelsinger and his then wife, Pattie, of Tucson, Arizona. (Pattie and Paul divorced a few years later.) The second of four children, Jesse was an apparently normal child until late in his third year, when his behavior and speech became erratic. "It seemed like demonic possession," Paul Gelsinger told me in a 2006 interview. "The voice coming out of him, the attitude, I thought it was some kind of psychiatric problem."

Eventually Jesse slipped into a coma, and this led to his hospitalization and his eventual diagnosis as having OTC deficiency. No one else among his relatives had had the disorder; the mutation apparently occurred spontaneously in one of Jesse's cells when he was a very early embryo. The descendents of that cell, but not those of the remaining embryonic cells, were OTC-deficient, making him a mosaic. Jesse's condition was so unusual that researchers at the University of Pennsylvania wrote an article about him that was published in the *New England Journal of Medicine* in 1988. Thus Jesse's case was well known to the community of specialists who studied and treated OTC deficiency, long before he became a subject in Wilson's clinical trial.

Jesse recovered from that episode, and thereafter he was maintained in reasonably good health with a combination of a low-protein diet and a drug, sodium benzoate, that lowered the concentration of ammonia in his blood. Still, the dietary restriction slowed his growth—he reached a final height of only 5'5"—and his metabolic problems affected his mental abilities to a variable extent. "When he was well, he was fine," his father told me. "Very intelligent—he could be an honor roll student. But at other times it was very difficult for him to focus."

In the fall of 1998, when Jesse was seventeen years old and in his final year in high school, he and his father received some interesting news. Jesse's geneticist, Randy Heidenreich of the University of Arizona, told them that he had received a letter from Mark Batshaw, a pediatrician and expert in OTC deficiency at the University of Pennsylvania. Batshaw had teamed up with James Wilson and a liver surgeon, Steven Raper, to run the first clinical trial of Wilson's adenovirus vector, and Batshaw was now actively recruiting volunteers. The Gelsingers reacted very positively, but the minimum age for participation was eighteen, so Jesse could not sign up for the trial until the following summer.

The intervening months were turbulent ones for Jesse and his family. Jesse had no plans for what to do after high school, aside from a wholly impractical dream of turning his favorite hobby—watching professional wrestling—into a career option. He had fantasies of starting his own pro wrestling federation. Tensions developed between Jesse and his father, who tried to focus his son's attention on the need to think about his future in a serious way, particularly because his medical condition involved considerable expenses—expenses that his father's health insurance would cover for only a few more years.

Jesse's normal teenage rebelliousness had a detrimental influence on his always-precarious health. "He consciously did not want to take his medication because of the peer effect," said Paul. "At school he would have to go to the nurse's office to take it. He was definitely different because of the disorder, and he hated that." Jesse began skipping some of the forty-odd pills that he had to take every day. Sometimes he would go without his medications altogether if he felt that he was well enough to do so.

Then, in November, Jesse camped out all night outside a box office with the hope of getting tickets for a pro wrestling event. A healthy teen's body would have taken such an overnighter in stride, but for Jesse it was the kind of stressful event that exacerbated his illness. He began experiencing serious symptoms of his disorder such as nausea and cognitive impairment, but he hid them from his father and from his stepmother, Mickie, in order to avoid having restrictions placed on his activities.

Three days before Christmas, Paul arrived home to find Jesse vomiting in the living room. He was admitted to the hospital, where tests revealed that his blood ammonia levels were six times above normal. His metabolism was falling into a vicious cycle whereby having consumed all the fat in his body, it was now digesting his proteins, thus liberating even more ammonia. Six days later, after a rocky course, he became delirious. Thinking that he was moribund, his doctors asked Paul whether he wanted a "do not resuscitate" order placed on his son. Paul vehemently refused. The doctors then decided to move Jesse into intensive care, but before they could do so Jesse stopped breathing. Luckily Paul was present at his bedside: he summoned a doctor who called a code blue. Jesse was incubated and put on a respirator.

A new and more powerful medicine—sodium phenylbutyrate—was fed to Jesse via a stomach tube, and it eventually lowered his ammonia level to the point where he regained consciousness and began a complete recovery. He returned to school with ammonia levels near normal for the first time in his life and with a newfound resolve to follow his doctors' orders. A few months later he graduated from high school.

On June 18, 1999—Jesse's eighteenth birthday—the entire family flew east. After a few days' visit with relatives in New Jersey, they drove down to Philadelphia so that Jesse could sign up for the OTC trial. The person who explained the trial and walked Jesse and Paul through the "informed consent form" was Steven Raper, the liver surgeon.

Raper said that the adenoviral vector would be infused directly into Jesse's hepatic artery—the artery that supplies the liver. The idea behind this was that most or all of the viral particles would be taken up by the liver, where the new gene was needed, and the rest of the body would be spared any ill-effects of infection by the virus. To reach the hepatic artery, Raper would have to insert a flexible cannula into the femoral artery in Jesse's groin and thread it backward up the aorta, in a similar fashion to what is done for coronary angiography. The infusion of the vector would take a few minutes, he said, and Jesse would have to lie still for several hours afterward. Over the following days blood tests would be done to check whether there was any effect on Jesse's ability to metabolize ammonia. Then, a week after the infusion, Raper would remove a small sample of Jesse's liver by means of a needle stuck through the front of his abdomen. This biopsy sample would be studied to test whether the vector had been taken up by the liver cells, whether the new gene was working, and whether the vector had caused any damage to the liver. Raper emphasized that although the infusion might cause some brief improvement in Jesse's ability to excrete ammonia, it would not offer him any long-term benefit. The benefit might come eventually to others, especially to OTC-deficient babies: The vector might help tide them over their first neonatal crisis and save them from immediate death or brain damage. In the long run, the hope was to develop other vectors that would implant the OTC gene more permanently in the children's genomes.

The informed consent form mentioned a laundry list of possible ill-effects that Jesse might suffer. It was quite possible that he would experience mild flulike symptoms over the day or so after the infusion. Blood clots might break loose. The vector might cause hepatitis. The needle biopsy might cause a hemorrhage. There might be unforeseen harmful consequences. Jesse signed the consent form, and he had blood drawn to measure his ammonia level. Then he underwent a several-hour test in which he had to drink a sample of ammonia labeled with a nonradioactive isotope of nitrogen (^{15}N). The fate of the nitrogen in this test would reveal how efficient Jesse's metabolism was at getting rid of ammonia.

Once out of the hospital, the family did a bit of sightseeing: They went over to the Spectrum Arena to see the famous statue of Sylvester Stallone as Rocky Balboa. A photo of Jesse standing in front of Rocky, his arms raised high in triumph, later accompanied many news stories about him, and it still can readily be found on the Internet. It aptly illustrated the new and positive focus that participation in the OTC trial had injected into Jesse's life—a life that otherwise was drifting rather aimlessly. After another day of sightseeing, this time in New York, the family returned to Tucson, where Jesse waited to hear more from the Penn team.

"Informed consent" is really a figure of speech. No layperson can truly evaluate the potential risks and benefits of participating in a clinical trial, least of all the trial of a genetically engineered virus. To some extent, signing a consent form is a confession of faith—faith that the researchers have done their homework and that the experimental protocol has been adequately reviewed by experts. In the case of Wilson's OTC trial, it looked like the review process had been extraordinarily thorough. Wilson's protocol for the study had been reviewed by the University of Pennsylvania's Institutional Review Board (IRB), the NIH, and the FDA. It had also undergone a special review by the FDA's Recombinant DNA Advisory Committee, or RAC—a group that vetted protocols involving genetically engineered viruses and other biological therapies.

Yet all had not gone smoothly during the approval process. For one thing, there had been setbacks during the animal testing that preceded the clinical trial. Three monkeys who had received very high doses of the vector developed severe liver failure combined with a blood-clotting disorder, and they had to be killed.

In reaction to these deaths Wilson prepared a second version of the vector that he claimed was safer. But was it? In 2006 I put this question to Inder Verma, a leading virologist and gene-therapy expert at San Diego's Salk Institute. "It's possible," Verma said, "but he had no proof of that. And in fact it's ironic, because we proved later on that *every* batch of adenovirus had that problem; it didn't matter whether it was the first, second, or third version. The viral proteins are going into cells and [causing them to be] recognized as foreign by cytotoxic T lymphocytes, which destroy them."

Some of the initial reviewers had serious reservations about the study. Among them was Robert Erickson, a pediatric geneticist at the University of Arizona who was a member of the RAC. In a 2006 interview Erickson told me that he had been concerned by the adverse events in the animal studies and also by Wilson's plan to infuse the vector into the hepatic artery, which Erickson viewed as a risky procedure. He changed his mind when Wilson described changes to the vector and also agreed to infuse it into a peripheral vein. (That change got reversed by a subsequent FDA panel.)

In the final plan for the trial, the vector would be administered to eighteen adult volunteers. The volunteers were to be in good health with their OTC deficiency under reasonable control, which meant their plasma ammonia concentrations could be no higher than 70 micromoles per liter (μM/L)—about twice the maximum level seen in healthy people. The volunteers would be divided into six cohorts, with three volunteers in each cohort. The volunteers in the first cohort would receive a tiny amount of the vector, and—assuming there were no ill-effects—the dose would be increased stepwise until the sixth cohort received the maximum dose. If there were serious adverse effects, they would have to be reported to the FDA before the trial could proceed further.

Another safety consideration had to do with the sex of the volunteers. Because women, with their two X chromosomes, are generally less severely affected by OTC deficiency than are men, it was decided that the first two volunteers in each cohort would be women. Men, if they participated at all, could only be the third and last in a cohort. In that way the doctors would already have some experience with that dose level before they treated a man. Jesse would therefore be the last of his cohort to receive the vector.

A few weeks after the visit to Philadelphia, Jesse and Paul got a letter from Mark Batshaw, the OTC deficiency specialist on the

Penn trial. Batshaw wrote that Jesse's test results made him an acceptable subject for the trial. His blood ammonia was below the cutoff level of 70 µM/L, and his efficiency at excreting ammonia, as measured by the ^{15}N test, was 6 percent of normal. Because of this very low efficiency—the lowest of anyone in the trial—any increase in ammonia excretion caused by the viral OTC gene would be readily apparent.

Soon after, Paul had a phone conversation with Batshaw, in which, according to Paul, Batshaw mentioned the good results they'd had to date. In experiments on OTC-deficient mice, he said, the gene transfer had worked so well that the mice had been protected from what would otherwise have been a lethal dose of ammonia. And because the human trial had now been under way for more than a year, Batshaw was able to give Paul some idea of the early results. The most recent of the volunteers, he said, had experienced a 50-percent increase in the efficiency of her ammonia excretion after the infusion of the vector. He did not mention any adverse events in the human subjects, according to Gelsinger. In a brief 2007 e-mail exchange, Batshaw confirmed the general content of the conversation but said that he remembered having told Gelsinger of adverse effects, including fever and short-term liver abnormalities. (Batshaw—like Wilson and Raper—declined my request for an interview.)

Both Paul and Jesse reacted with enthusiasm to this news, and Jesse was doubly excited about participating. Because the trial was now nearing its end, Jesse would be in the final cohort, and he would therefore receive the highest dose of the vector. But he would have to cool his heels for a while, because as a male he would be the last of the three volunteers in that group. Thus he would be the final subject in the entire trial. His infusion was scheduled for October.

Jesse spent the intervening time working as a supermarket clerk and, in his free hours, riding a motorbike that his father and stepmother had given him as a graduation present. He seemed as upbeat and full of life as Paul had ever known him.

In mid-August the coordinator of the clinical trial called to say that their next patient (the second patient in the last cohort) had a scheduling conflict, and that they would like Jesse to take her place.

This would mean that the infusion would take place in September. Jesse agreed to the change of date.

Putting Jesse—a male—into the second position in a cohort was a clear-cut violation of the protocol that Wilson and his colleagues had agreed to. Wilson has never denied this, although public statements put out by the Institute of Gene Therapy have maintained that the FDA OK'd a similar switch in an earlier cohort, so Wilson felt entitled to make the switch in this cohort too, even without express permission.

According to Paul Gelsinger, the reason for the switch was not that the other patient had a problem with the September date, but that she backed out of the trial altogether. (I was not able to ask any of the investigators directly about this, however.) If the other patient did drop out, it might have seemed best to infuse Jesse right away and wrap up the trial rather than endure the delays involved in recruiting another volunteer. "They just wanted to finish," Inder Verma speculated, "because this was the last dose and they wanted to stop and get it over with."

Unknown to Jesse or his father, a number of untoward events occurred at Penn prior to Jesse's visit. Several of the earlier volunteers experienced significant liver damage from the adenovirus infusions. The damage was assessed by measuring enzymes, such as transaminase, that were released from dying liver cells into the bloodstream. Even as early as June of 1998 one of the volunteers who had just been infused with the adenovirus vector experienced a surge in her serum transaminase levels to nearly eight times the upper limit of normal.

This finding indicated very significant damage to the woman's liver, even at a dose of the vector far below that which Jesse was scheduled to receive. It was a "grade-III adverse event" on the scale of severity established by the FDA, but Wilson reported it to the FDA as a milder "grade-II" event. Another grade-III event (a high fever) was also reported as grade-II. Numerous other volunteers experienced grade-II adverse events—in fact there were grade-II or grade-III events in every cohort from the second one onward. According to Paul Gelsinger, all four of the volunteers who immediately preceded

Jesse experienced grade-III liver toxicity, but I could not find independent documentation of this. Nevertheless, an official letter of reprimand later sent to Batshaw by the FDA listed five grade-III adverse events, two involving liver toxicity and three involving high fever.

The protocol called for halting the trial for regulatory review if there was a single grade-III event or at least two grade-II events, but the Penn researchers did not halt the trial. They reported the adverse events many months after they occurred, if they reported them at all, and often the information was hidden in the back pages of their reports, while the summaries at the front were much more positive. In the cover letter to his IRB report of August 9, 1999, for example, Wilson wrote as follows: "No serious adverse effects have occurred as a result of this study. There have been no significant treatment-related toxicities or procedure-related toxicities, and all participants have remained well." Much later, the usually circumspect FDA described this statement with a simple and damning adjective: "false."

Jesse Gelsinger took an unpaid leave from his job, and on Thursday, September 9, he flew alone to Philadelphia, taking with him a bag of clothing and his collection of pro wrestling videos. The plan was that Paul, who could not take a great deal of time off from his work, would join Jesse a week later for the liver biopsy, which Paul perceived to be the most risky part of the trial. "Words cannot express how proud I was of this kid," Paul wrote later. "Just eighteen, he was going off to help the world. As I walked him to his gate I gave him a big hug and as I looked him in the eye, I told him he was my hero."

Jesse checked into the hospital on the Thursday evening. The next few days would be devoted to tests, and the adenovirus infusion was scheduled for Monday. But almost immediately a problem arose: On Friday morning, before Jesse underwent a ^{15}N ammonia excretion test, his blood ammonia level was 114 µM/L—well above the permissible maximum of 70 µM/L. Batshaw and Raper gave Jesse intravenous drugs to lower his ammonia, but by Sunday, the day before the scheduled treatment, it had only fallen to 91 µM/L. This presented a final opportunity—and an obligation, according to the protocol—for the research team to drop Jesse from the trial, but they did not do so.

On Monday morning Steven Raper inserted the infusion cannula into Jesse's groin and threaded it up into the hepatic artery. Having checked the correct placement of the cannula tip with the help of radiography, Raper began the infusion of the adenoviral vector. A total of 38 trillion virus particles entered Jesse's bloodstream over the course of a few minutes. When the infusion was over, Jesse had to lie quietly for a few hours and during this time Raper called Paul to let him know that everything had gone according to plan. That evening, Jesse developed a fever of 104.5°F, but this didn't prevent Jesse from talking with his family by phone—for the last time, as it turned out.

The next morning, Raper noticed that the whites of Jesse's eyes were yellow: a sign of jaundice, which suggested damage to his liver or to his red blood cells—or both. Jesse was also slightly disoriented. Raper called Paul to let him know, and he also talked with Mark Batshaw, who was in Washington. Later, Batshaw called Paul to tell him that Jesse's blood ammonia level had risen to 250 µM/L, and that he was seriously ill. Paul dropped everything and took a night flight to Philadelphia. When he arrived at the hospital on Wednesday morning, he found Jesse on a ventilator and in a coma. His blood ammonia level had risen at one point to 393 µM/L—about ten times higher than normal—but the doctors had been able to lower it to less than 70 µM/L by putting Jesse on dialysis. Still, Jesse remained desperately ill: In particular, his blood was being poorly oxygenated in spite of the ventilator. In addition, he was developing disseminated intravascular coagulation, a potentially deadly condition in which the blood clots inside the blood vessels. Jesse didn't respond to any of Paul's efforts to rouse him.

Paul called his wife and told her to come join him. Some of Paul's many siblings also converged on the hospital. For a while it looked as if Jesse was improving, but by late Wednesday evening it was apparent that his lungs were failing. Even while being mechanically ventilated with pure oxygen his arterial blood contained insufficient oxygen to maintain his vital organs.

In a last-ditch effort to save Jesse's life Raper decided to try hook-ing Jesse up to an artificial lung—a device resembling the heart-lung machines that are used during open-heart surgery. It took Raper and the machine specialist until five a.m. on Thursday to get the machine hooked up and running—there was major bleeding, and they had to use more than ten units of blood to make up the deficit and prime the machine. Paul spent the night waiting in quiet desperation outside the intensive care unit, with an occasional brief visit from a harried doctor and some comfort from the hospital chaplain.

Meanwhile, the elements were conspiring to echo Jesse's crisis: Hurricane Floyd was grinding its way up the East Coast toward Pennsylvania. Paul's wife, Mickie, made it to Philadelphia just before the airport was closed, but Mark Batshaw became trapped on a train that was stuck outside the city. He ended up giving advice by cell phone from the train.

For a while, it seemed as if the artificial lung was helping. Sometime after noon, Paul, Mickie, and several of Paul's siblings were allowed in to visit Jesse. He was deeply comatose, and his face and body were enormously bloated. Paul was not even sure that it was his son he was looking at, until he spotted a familiar tattoo and scar.

Toward evening, Paul and Mickie returned to their hotel, but Paul was destined for yet another sleepless night. At one point he walked back to the hospital through the rain, only to find Jesse in even worse shape: He was losing blood in his urine.

When Paul and Mickie came back in to the hospital on Friday morning, Raper and Batshaw told them that Jesse's brain had suffered irreversible damage and his other organs were failing too. They suggested that it was time to shut off his life support. For the final ceremony, seven of Paul's siblings and their spouses joined Paul and Mickie and about ten of the hospital staff around Jesse's bed. After a prayer from the chaplain and brief words from Paul, Raper clamped off the tube that carried blood to the artificial lung and switched off the machine. A minute later he put his stethoscope to Jesse's chest. "Good-bye, Jesse," he said. "We'll figure this out."

In the immediate aftermath of Jesse's death, Paul was not inclined to blame the researchers. "It was a traumatic experience for me," Paul told me, "but it was also a spiritual experience. Jesse's example, what he was doing, demonstrated the best of humanity to me—the best that we can be. At the time I was adopting that attitude: forgiveness, honesty, everything—the best that we can be." Yes, participating in the clinical trial had killed Jesse, but it seemed to Paul that it was an unforeseeable accident. Paul told Batshaw as much, and said that he would not be bringing a lawsuit.

In the same spirit, Paul invited Steven Raper to attend the scattering of Jesse's ashes. That ceremony took place at one of Jesse's favorite spots, the 9,400-foot summit of Mt. Wrightson, thirty miles south of Tucson. Jesse's mother, Pattie, who had been ill at the time of Jesse's death, did accompany the remainder of the family on the arduous hike. On the descent after the ceremony Pattie fell behind the rest of the party; she was overtaken by darkness and had to be rescued.

Paul's forgiving attitude changed after he was invited to address a meeting of the Recombinant DNA Advisory Committee that would discuss Jesse's death and what it meant for the regulation of gene-therapy experiments. The meeting was to be held in Washington, D.C., in early December of 1999. In the days before his trip Paul began to learn facts about the case, such as Jesse's high ammonia level before the infusion, which made him think that the FDA had been lax in their oversight of the trial. He said as much to an FDA staff member and even threatened to expose what he saw as the FDA's dereliction of duty.

Then James Wilson asked Paul Gelsinger to come to Penn a day before the RAC meeting in order to give a "morale boost" to his institute's deeply shaken staff. When Paul arrived, Wilson took him into his office. "He'd been all enthusiastic about my visit until that point," Paul told me, "but now he was crying, and he said that he'd just received notice, a press release [from the FDA], and they were pointing at him and his colleagues as being responsible for Jesse's death." As Paul sees it, the FDA had taken fright at Paul's accusations and had quickly moved to shift blame onto Wilson, Batshaw, and Raper. It's equally possible, however, that the FDA had simply reached this conclusion on the basis of its own weeks-long analysis of the case. (Kathryn Zoon, the responsible FDA official at the time, declined my request for an interview.)

The RAC meeting was thrown into turmoil by the FDA's announcement: Throngs of reporters and lawyers almost outnumbered the scientists and regulators. Paul Gelsinger delivered a long presentation describing his entire experience with the OTC trial. He was still not publicly blaming the Penn group for his son's death. He told a reporter for the *New York Times* that "These guys didn't do anything wrong." But his mindset was rapidly changing. For one thing, Paul told me that he found out at the meeting that there was no evidence for any improvement in ammonia excretion among the volunteers who received infusions of the adenovirus vector before Jesse. This conflicted with what he says Batshaw told him before Jesse's participation, which was that the volunteer immediately before Jesse had experienced a 50-percent improvement in ammonia excretion. Paul began to feel that Batshaw had deceived him about this in order to make him feel that the trial was going well, when it wasn't.

Over the following months Paul probed deeper and deeper into the events preceding his son's death, and he became convinced that there had been serious wrongdoing on the part of the Penn researchers. Besides the problems already mentioned—the failure to halt the trial when previous volunteers experienced serious adverse events and the decision to go ahead with Jesse's infusion when his ammonia was above the allowable limits—Paul learned that James Wilson had what he considered to be a financial conflict of interest.

The details of Wilson's financial involvements were unraveled by *Washington Post* reporters Deborah Nelson and Rick Weiss in November of 1999. Wilson owned 30 percent of the stock of a biotech company named Genovo, which he had founded to commercially exploit gene therapies. This company was paying 20 percent of the operating expenses of the Institute for Human Gene Therapy in return for the right to exploit the institute's discoveries. The University of Pennsylvania also had financial ties to Genovo and a web of linked biotech corporations. How could Wilson give priority to the safety of the volunteers, Paul reasoned, when his company stood to profit from quick results?

In fact, just a few months before Jesse's death Wilson had delivered a presidential address to the American Society of Gene Therapy in which he recommended a streamlining of the regulatory process. The initial or phase-1 trials, he said, should be explicitly designed to gain information on the *efficacy* of gene-transfer therapies, not just on their safety as had been the traditional purpose of phase-1 trials. "Early feedback on clinical or surrogate measures of efficacy can have broad implications in accelerating the path to commercialization," he said, "focusing our investment in research, and minimizing financial risks regarding decisions to move forward in later stage development." This sounded like the opinion of an investor, not of a scientist or doctor.

The University of Pennsylvania commissioned an independent panel to review the Institute of Human Gene Therapy in the wake of Jesse's death. The panel, one of whose members was Inder Verma, recommended that clinical investigators who were testing biological therapies not be allowed to have investments in companies that were commercializing such therapies. This seemed like a rebuke to Wilson for his financial interest in Genovo, but when I asked Verma about it in 2006, he denied that. Genovo, he said, wasn't developing the same kind of vector that Wilson's team tested on Jesse, so there was no financial conflict of interest. "The newspapers liked to say that there was—it was so inflammatory—but it was a different virus." Robert Erickson, on the other hand, did see a clear conflict of interest in Wilson's relationship with Genovo. He argued that because the two vectors fell under the same umbrella of gene therapy, and because Genovo had broad rights to the Institute's discoveries, Genovo and therefore Wilson stood to gain financially if the OTC trial got good results.

Yet another view of the matter was offered by Arthur Caplan, the Penn ethicist. "I think the problem was ambition, not money," he told me. "The problem was the drive to succeed, or to be the first to show something efficacious with gene therapy. Batshaw and Wilson, when they stood up in front of their peers, they were very interested in saying, 'We at Penn are the very first ones to make this much-hyped idea pay off.' To the extent that they hurried or didn't worry about signals from the animals that there were problems, that was the reason. And we haven't figured out how to manage *that* conflict of interest."

One year to the day after Jesse's death, Paul and Jesse's uncle (the administrator of Jesse's estate) filed a lawsuit against the University of Pennsylvania, the Children's National Medical Center (Batshaw's institution), the Genovo Corporation, and five individuals: Wilson, Batshaw, and Raper, along with the dean of the Medical School and Arthur Caplan. The lawsuit alleged wrongful death, fraud, and other misdeeds.

Caplan was soon dropped from the suit on account of his peripheral involvement. The remaining defendants settled with the Gelsingers within six weeks of the filing. The amount of the settlement was never disclosed, but it was clearly very substantial: Paul no longer works as a handyman, and he mentioned to me that Pattie's share of the settlement has allowed her to get good medical care for the first time in her life. Paul did not get an apology or even an admission of wrongdoing, however: The University's position was that, while there had been lapses in the oversight and execution of Wilson's trial, these were not what led to Jesse's death.

In mid-2002, after critical assessments by the independent review panel, Wilson resigned as director of the Institute for Human Gene Therapy. Soon thereafter the institute itself was closed down.

The Gelsinger lawsuit and the closure of the institute were not the only woes that Wilson and his colleagues faced, however, because the FDA also sought retribution on the Penn team, and it did so with a single-minded focus and almost nitpicking attention to detail that far outdid its performance while regulating the clinical trial itself. Over months and years, long lists of allegations flew from Washington to Philadelphia, and even longer explanations and refutations flew back from Philadelphia to Washington. The core of the government's case was that Wilson and his coinvestigators had failed to stop the clinical trial or to properly notify the FDA when earlier participants in the trial developed severe adverse reactions to the infusion of the adenoviral vector, and that they proceeded with the infusion of Jesse Gelsinger when his high ammonia levels should have disqualified him from participation. Absent these misdeeds, Jesse would not have died.

Eventually the FDA turned the matter over to the U.S. Department of Justice. The DOJ brought a formal action against Wilson, his colleagues, and their institutions, alleging that they had committed numerous violations of the Civil False Claims Act in their dealings with the NIH and the FDA.

Finally, in February of 2005, a settlement was reached. Under the terms of the settlement, none of the defendants admitted to the government's allegations, but the University of Pennsylvania and the Children's National Medical Center each agreed to pay a penalty of more than $500,000 and to enact numerous changes to strengthen the oversight of clinical trials. Wilson was banned from participating in clinical trials until 2010, and a special monitor was to be appointed to oversee his other research activities. He would have to undergo retraining if he ever planned to participate in clinical trials again. In addition, he was to write and publish an article describing the "lessons learned" from the entire episode. Batshaw and Raper were placed under government supervision for a period of three years and would have to undergo retraining.

These were severe punishments for ambitious clinical researchers, but Paul Gelsinger doesn't see it that way. "They got off easy, amazingly easy," he said to me. "It was medical manslaughter." The fact that Wilson never apologized or admitted any culpability for Jesse's death was particularly galling to him.

With Batshaw and Raper it was a little different. Whereas Paul never met Wilson until after his son's death, he had established some rapport with both Batshaw and Raper and had witnessed their efforts to save Jesse's life. Although they may not have explicitly apologized or admitted culpability, Paul read their actions as expressing a desire for forgiveness. Raper, after all, climbed 5,000 vertical feet with Jesse's family to scatter a portion of his ashes. And Batshaw, about a year after the settlement, sent Paul an e-mail in which he said that his thoughts had turned to Jesse on the occasion of Yom Kippur, the Jewish Day of Atonement. Later the two men met briefly. Paul acknowledged to me that both Batshaw and Raper were genuinely torn up by what happened. "I'd like to see these guys find forgiveness for what they did," Paul told me. "But part of that is acknowledging what you've done."

Did the tragedy of Jesse's death, the ensuing publicity, the investigations, the legal maneuvers, and the ensuing tightening of regulations improve the safety of clinical trials? To Robert Erickson, the University of Arizona geneticist who reviewed Wilson's initial proposal, the answer is yes. He said that, after it was revealed that Wilson had failed to report his volunteers' adverse reactions, seven hundred such reports came in from other clinical investigators within a period of just a few days. "I think people are reporting adverse reactions much more quickly," he said.

Arthur Caplan takes a less positive view. "Paul Gelsinger tried to bring about changes so that the same thing wouldn't happen to anyone else," he told me. "He went to a lot of meetings, he fought for subject rights, he gave of himself in a very powerful way, he became committed to trying to make change. But you know what I think the legacy of Jesse's death was for human subjects reform? Almost nothing. Many of the problems that existed then still exist now: failure of informed consent to work well, IRBs still overwhelmed with work, no clarity on the animal data reports, difficulties in getting subject selection criteria properly understood. I would say no, it's no better." And he went through a litany of clinical trials that have gone seriously wrong since Jesse's death.

One particularly nightmarish episode occurred in 2006—not in the United States, but in Britain. In March of that year a drug-testing company named Parexel conducted the initial human trial of the immune-system modulating drug TGN1412, which had been developed by a German biotech company named TeGenero. The drug was a monoclonal antibody—an immune-system molecule that binds to a single molecular target in the body. The antibody was designed to stimulate the class of white blood cells known as T cells, and the hope was to utilize this effect in the treatment for rheumatoid arthritis and a certain form of leukemia.

After uneventful trials in laboratory animals, Parexel recruited eight healthy volunteers—all men—for the initial human trial, which was to be conducted at Parexel's unit at Northwick Park Hospital in North London. The volunteers included Navneet Modi, a twenty-four-year-old business-school graduate, Ryan Wilson, a twenty-year-old trainee plumber, and Mohammed Abdalla, a twenty-eight-year-old bar manager. Each of the eight subjects was to receive £2,000 ($3,800), a sum well above what is customary for participation in drug trials in the United Kingdom. In fact, some of the men were basically professional guinea-pigs. They derived a significant portion of their income from volunteering for drug trials. (Jesse Gelsinger, in contrast, was never offered any financial incentive to participate in the Penn trial.)

Six of the subjects were infused with the drug at a dosage of 0.1mg/kg. This level was considered safe because monkeys had tolerated dosages five hundred times higher without any apparent ill-effects. Two other subjects received an inactive placebo. Each infusion took about two minutes, and the eight infusions were done one immediately after the other, so that the whole procedure took about twenty minutes.

"An hour after the drug entered my body I was suddenly gripped by pain," Modi later told *The Times* (London). "I felt my head swelling up like an elephant's. I thought my eyeballs were going to pop out. I screamed out, 'Please, doctor, help me. Help me,' but he told me to lie down, then came back with a single paracetamol [Tylenol] tablet. It felt like a terrible nightmare."

All six of the subjects who received the TGN1412 became extremely ill. They were experiencing the same kind of "cytokine storm" that had killed Jesse Gelsinger. Their heads and bodies grotesquely swollen, they were rushed into intensive care. Within twelve hours their lungs, kidneys, and other organs were starting to fail, and their blood pressure fell to dangerous levels. The subjects were treated with massive doses of steroids and other immune-suppressing drugs, and they underwent dialysis in an attempt to remove the TGN1412 from their bodies.

Four of the men began to recover within two days, but two men—Wilson and Abdalla—became critically ill with cardiovascular shock and respiratory distress syndrome, and were maintained on respirators and other life-support systems for many days. Eventually all of the subjects recovered sufficiently to be discharged from the hospital, but they have had serious ongoing medical problems, including blackouts and diarrhea. Wilson had to have gangrenous toes and fingertips amputated. All the subjects are considered to remain at high risk of cancer and immune-system disturbances; in fact, one of them was already reported to show early signs of lymphoid cancer four months after the drug trial. "It's a really bizarre feeling when you discover you might be dead in a couple of years or even in a couple of months," said Modi. "I feel like I've given away my life for £2,000."

Initial investigations into the cause of the medical disaster have failed to identify any human error, such as in dosage, that might have caused the drug trial to go so seriously wrong. Nor does it seem that the drug was contaminated with bacteria or with other toxic agents. Rather, it appears that this was an unforeseen reaction to TGN1412 in its intended dosage. Nevertheless, two aspects of the way the trial was conducted seem to have played key roles in causing the harm to be so severe. First, and most obviously, the researchers did not wait to observe the reaction of the first volunteer to the drug before infusing it into the other five. "To me, that was insane," commented Robert Erickson. Second, the researchers infused the drug at a very high rate—much faster than it had been infused into the monkeys. Adverse reactions are typically worsened by high infusion rates, so a slower rate might have caused lesser harm or none at all. Nevertheless, the investigation when complete may identify other more important causal factors.

Of course, a lawsuit is now in the works. But TeGenero, a recent biotech start-up, filed for bankruptcy in July 2006, citing the drying-up of investment capital after the drug-trial story broke. Each subject reportedly received £10,000 in compensation from TeGenero before it went bankrupt, but the company's insurance policy was for only £2 million, meaning that there will not be nearly enough money to adequately recompense the six men should the company be found liable.

———

Gene therapy took a terrible blow from the Gelsinger tragedy. Clinical trials came to a halt and were only restarted in the most restrictive and cautious fashion. In 2006 the "father" of the field, USC's William French Anderson, was convicted of child molestation and (in the following year) sentenced to fourteen years' imprisonment. Then, in July of 2007, another death occurred. This happened during a clinical trial of a genetically engineered vector designed to treat rheumatoid arthritis and related conditions. The vector, developed by a Seattle-based company named Targeted Genetics, employed a type of virus different from (and supposedly safer than) the adenovirus that killed Jesse Gelsinger. The trial was also thought to be safer because the vector was injected locally into the affected joint, rather than into the bloodstream. Yet the patient died after receiving a second dose of the vector. The patient's identity and the exact circumstances of his or her death have not yet been disclosed.

In spite of these setbacks, Inder Verma sees a bright future for the field. He points to a dozen or so SCID-affected children who are alive today because of gene-therapy procedures that they underwent in France. "It will become a successful therapy," he says.

CHAPTER 7

⚙

NUCLEAR PHYSICS:
Meltdown

The tombstone of Richard Leroy McKinley looks no different from hundreds of others in Section 31 of Arlington National Cemetery. It's a plain stone slab, decorated with a simple cross. It lists McKinley's date of birth (December 2, 1933) and death (January 3, 1961), states his rank (Army Specialist 4), and mentions his service in Korea.

If you were to dig beneath the stone, however—an act which is forbidden by a special order from the office of the U.S. Adjutant General—you would find something out of the ordinary. You would have to dig through three feet of earth, then drill through a foot-thick slab of concrete, then break open a metal enclosure that reaches ten feet into the ground, and then work your way through another concrete casing before you got to the metal casket, which is lined with lead sheeting. If you finally managed to get the casket open, you would find what appeared to be a mummy, wrapped in successive layers of lead, plastic, and cotton sheeting. After unwinding these wrappings, you would finally see the mortal remains of Richard McKinley himself. You would notice that his belly and chest have been roughly sliced open and his internal organs removed, along with his left arm and most of his skin. While pondering this macabre scene, you would be absorbing enough nuclear radiation to put your life in peril.

McKinley was one of three men—the others were John Byrnes and Richard Legg—who died in a nighttime explosion at the National Reactor Testing Station (now the Idaho National Laboratory) on Idaho's Snake River Plain. The cause of the explosion, in a scientific sense, was quickly figured out. But the human cause of the accident—if indeed it was an accident—remains a mystery forty-five years after the event.

The Testing Station opened in 1949. Throughout the 1950s and beyond, it was ground central for America's effort to develop controlled nuclear fission as the basis for power generation, both for military and peaceful applications. The Station's first reactor went critical in 1951, and a couple dozen more were constructed during the remainder of the decade.

The reactor where the three men died became operational in 1958. It was called Stationary Low-Power Reactor Unit 1, or SL-1 for short. The "stationary" in its name set it apart from the reactors then being developed for submarines, ships, and even aircraft. The "low-power" referred to the fact that SL-1 was designed as a compact, easily transported, and easily operated device that could provide modest amounts of electricity and heat. Its main intended application was to power the stations of the DEWLine—the radar and radio installations that had been strung out across the North American Arctic to provide early warning of a Soviet bomber or missile attack.

The SL-1 reactor was located, like all the other Testing Station reactors, on a sagebrush desert in the middle of nowhere. The nearest sizeable community, the town of Idaho Falls, lay forty miles to the east. That was where the reactor operators lived. Because of the lack of a nearby population that might be put at risk by a nuclear accident, and because the reactor was designed for use in even more remote locations, it had no concrete containment structure. Instead, it was housed in a simple silo-like building, about fifty feet high and clad in sheet metal.

The building had three levels. In the lowest level sat the reactor itself, buried in a bed of gravel. The middle level—the reactor room—gave the operators access to the top of the reactor and the all-important control rods, as well as to the turbine generator that converted the reactor's heat to electricity. The top level (the "attic") contained a condenser and fans to convert the steam that exited the generator back to water. In addition, a long, low building directly abutted the silo. This contained the control room and offices.

The design of the reactor was quite simple. At the bottom of a fifteen-foot-high steel vessel was the reactor core, a set of fuel plates containing uranium metal that had been moderately enriched in the radioactive isotope uranium-235 or ^{235}U. The nucleus of a ^{235}U atom will disintegrate when it absorbs a low-energy neutron, and the disintegration is accompanied by the release of several high-energy neutrons. For a nuclear chain reaction to occur, the energy of the emitted neutrons must be reduced to a level that can trigger further disintegrations. In the SL-1 reactor this was accomplished by water that circulated between the fuel plates. Thus SL-1 was a "water-moderated" reactor. Water also served to carry heat from the reactor to the generator.

The essential requirement for a nuclear reactor is that the nuclear chain reaction be controllable. For that purpose, SL-1 was equipped with five control rods whose lowermost portions were blades made of cadmium metal, an efficient absorber of neutrons. Each rod weighed about one hundred pounds and was seven feet long. When the rods were all dropped to their lowest position so that the blades lay between the fuel plates, they absorbed enough neutrons to terminate the chain reaction. Raising the rods by a few inches allowed the reaction to begin. During normal operation the control rods were raised and lowered by motors located in the reactor room above the reactor, and the motors were controlled remotely by operators in the control room.

Jack Byrnes and Dick Legg joined the SL-1 staff in the fall of 1959. Byrnes was twenty years old; he came from Utica, New York, and he brought with him his nineteen-year-old wife, Arlene, and their one-year-old son. Legg was a single man from rural Michigan, but he married a woman from the local Mormon community soon after he arrived in Idaho Falls. Byrnes had enlisted in the Army, and Legg in the Navy, but the two men knew each other before they arrived in Idaho, because they had taken the same training program in Virginia over the previous months.

By the time of the accident on January 3, 1961, Byrnes and Legg had more than a year of experience operating the SL-1 reactor. Legg, in fact, had already been designated a chief operator and shift supervisor. Richard McKinley, on the other hand, was a new arrival and was still a trainee. He had much more experience in the military than the other two men, however; he was twenty-eight, and he had served both in the Army and the Air Force. He was married with two children.

On December 23, 1960, eleven days before the accident, SL-1 workers shut the reactor down for the Christmas holiday. They did this, as usual, by dropping the five control rods to their lowermost position. Several of the rods failed to drop under their own weight, however, and had to be driven down. This "sticking" behavior had been noticed before, but it wasn't considered a serious problem. The key central rod which, on account of its position, was capable of starting and stopping the nuclear chain reaction by itself, had never stuck, nor did it do so on this occasion.

During the day on January 3, a work crew detached the control rods from the motors that normally raised and lowered them, in order to conduct some tests. The workers left the task of reconnecting the rods and restarting the reactor to the night crew, which consisted of Byrnes, Legg, and McKinley. The three men came on duty in the late afternoon. The last contact that anyone in the outside world had with the crew was around seven p.m., when Byrnes's wife spoke with him by telephone.

At 9:01 p.m., an automatic alarm sounded at the Testing Station's firehouse, signaling a problem at the SL-1 reactor. There had been two such alarms earlier that day, and both had turned out to be false: They had been triggered by a faulty fire detector. Thus the firemen suspected that this alarm was false too. Nevertheless they set out in a car and fire truck, reaching the reactor at 9:10. When they entered the control room, they found it deserted, but radiation alarms were sounding, and a radiation detector carried by one of the firemen confirmed potentially dangerous levels of radiation.

The fire crew decided to check the reactor room, which was reached by a metal staircase that wound around the outside of the reactor building. By the time they were halfway up the staircase, their radiation detector was pegging out at 500 roentgens per hour, meaning that the men would absorb one year's permissible dose of radiation in less than two minutes. They retreated. Then one of them climbed quickly to the top of the staircase and took a glance inside the reactor room before descending again. He saw that the floor of the room was awash with water from the reactor and was littered with broken equipment and gravel from the shielding bed. He didn't see any of the three missing men.

A major disaster alarm was now sounded, and officials were roused at the Testing Station, in Idaho Falls, and even in Washington, D.C. Among the men who raced to the site was Ed Vallario, the health physicist responsible for the SL-1 reactor. At around 10:30, wearing respirators but no other protective gear, Vallario and another man ran up the staircase to the reactor room. Again, their radiation detectors went off scale. When they reached the entrance to the reactor room and looked inside, they saw Richard McKinley. He was lying by the control panel, moaning and twisting his body in an apparent attempt to reach the controls. The right side of his face was severely injured. Near him lay the lifeless body of Jack Byrnes. Dick Legg was nowhere to be seen.

The two rescuers beat a hasty retreat. Vallario knew that McKinley must have absorbed a lethal dose of radiation, but he was still alive, so he had to be rescued. Vallario and four other men found a stretcher and made another foray into the reactor room. They were able to get McKinley onto the stretcher and out of the building, but not before two of the men's respirators malfunctioned, forcing them to remove the respirators momentarily and to inhale the unfiltered, contaminated air of the reactor room.

The rescuers placed McKinley in a truck and then transferred him to an arriving ambulance. In the ambulance he was tended to by a nurse, Hazel Leisen, but within moments—before the ambulance could even get onto the main road—he died. Health physicists quickly pulled Leisen out of the ambulance, for radiation levels inside the vehicle were 500 roentgens per hour—enough to kill anyone who drove in it for more than a few minutes. The driver was wrapped in a lead blanket and told to drive the ambulance off into the sagebrush, so as not to expose people passing on the road. He did so, and then made a run for it.

With McKinley and Byrnes both dead, there was but one urgent question remaining: Where was the third member of the night crew? To find the answer, another team of four volunteers entered the reactor room. They searched around futilely and in increasing desperation as the seconds ticked away. Then one of the men's flashlights happened to point upward, and the men saw Dick Legg's body. It was pinned to the ceiling by one of the reactor's control rods, as if by a giant thumbtack. Evidently the control rod had shot explosively out of the reactor and had passed through Legg's pelvis and abdomen on the way to the ceiling, carrying his entire body with it. Legg's body now dangled limply and was very obviously dead. The men ran from the room.

An hour or two after midnight, teams went to the homes of the three dead men to notify their wives.

The acute emergency was now over, but two difficult problems lay ahead. The dead men needed to be removed from the site, decontaminated, autopsied, and buried. And the reactor needed to be rendered safe, broken down, and analyzed so that the cause of the accident could be established.

Richard McKinley's body needed to be dealt with first, because it was still lying in the abandoned ambulance out amongst the sagebrush. Operating in minute-long shifts, workers approached the ambulance and, using poles and other tools, succeeded in removing McKinley's clothes and wrapping the body in lead blankets. After roads were closed to other traffic, a volunteer put on a lead-lined jacket and drove the ambulance at high speed to the Testing Station's uranium reprocessing plant, where the body was placed in a steel bath containing ice and alcohol. The hope was that this would leach radioactive contamination off the body. Later that day, other teams of workers dashed into the reactor room and, working in relays, removed the body of Jack Byrnes, which was then also brought to the reprocessing plant.

Removing Dick Legg's body, impaled as it was to the ceiling of the reactor room, presented a far more challenging task. Not only was it in an extremely high-radiation environment; there was also the risk that if it fell onto the broken and exposed reactor during the recovery effort, it might trigger a second nuclear accident. At least the rescuers had time on their side, for the radiation was so intense that Legg's body had been completely sterilized: every bacterium inside and outside of his body had been killed, and the body would therefore not decompose.

As a first step in the recovery, radiation specialists built a full-scale mock-up of the reactor room, where they could test various strategies for reaching and removing the body. By five days after the accident they had settled on a strategy and put it into effect. First, they opened a large freight door that gave access to the reactor room from outside the building. Then they obtained a crane with a long boom and attached a netlike structure to the end of it. The crane operator, working blind behind a lead shield but guided by a distant spotter, maneuvered the net through the cargo door and under Legg's body. Then volunteers, working in minute-long relays, ran into the reactor room with long hooked poles and attempted to drag the body down from the control rod. After several relays, the mangled body did indeed fall into the net, and it was removed from the building and taken to the reprocessing plant.

In an official film made to document the SL-1 accident and its aftermath, the narrator stated that the three bodies were successfully decontaminated by washing, and thus were brought into a state in which they could be safely returned to their families and buried. This was a complete fiction, however. In reality, repeated washing, and even shaving of the men's body hair, had little effect in reducing radiation levels near the bodies, which still ranged up to 1,500 roentgens per hour. Countless particles of nuclear fuel had penetrated all three men's bodies, in effect converting them into high-level nuclear waste. No one could safely remain close to the bodies for more than a minute or so. Performing autopsies and dealing with the ultimate disposal of the bodies therefore presented an extraordinary challenge.

Three physicians from Los Alamos National Laboratory took on the task. Working behind lead shields set up several feet away, the doctors manipulated the bodies with instruments attached to long steel pipes. They first documented the injuries caused by the blast. Richard McKinley, the only one of the crew who was not killed instantly, had major wounds to his scalp, face, eyes, and to his left hand and left leg. Jack Byrnes, whose dead body had been found lying next to the reactor, had more severe injuries. His face, throat, rib cage, left arm, left leg, and back were completely crushed, and his pelvis had been driven up into his abdomen. As for Dick Legg, who had been found impaled into the ceiling of the reactor room, his body had suffered devastating damage. The top off his head had been sliced off, exposing his brain, and his face was collapsed inward. The upper half of his body had been twisted by 180 degrees with respect to the lower half, and his internal organs had been destroyed or displaced by the control rod as it blasted through his pelvis and abdomen. His left leg was cut almost in two.

With the clumsy tools available to them, the doctors cut open the bodies and removed the internal organs for microscopic study. A more difficult task was to reduce the radiation levels to a point that would permit the bodies to receive normal burial. Guided by radiation detectors, the doctors sliced away great swaths of skin and underlying tissue, but that wasn't enough. With McKinley, the doctors had to remove his left arm. With Byrnes, they removed both legs. With Dick Legg they had to go even further: They removed all four limbs and—after consultation with their superiors—his head, which was the most radioactive body part of all. All these removed body parts were packed in a drum, driven to a remote spot, and dropped into a deep trench designated to receive high-level waste from the accident.

The remaining portions of the men's bodies—just a limbless, headless torso in Legg's case—were now prepared for burial. The bodies were coated in drying powder, wrapped in cotton, and then in plastic sheeting. A final wrapping was done with lead sheeting—one-eighth of an inch think for McKinley and Byrnes, three-quarters of an inch for Legg. More than a ton of lead was used for this purpose. Thus ensheathed, the bodies were placed in hermetically sealed steel caskets that had been lined with more lead, and the caskets were placed within purpose-made lead vaults. Radiation levels outside the vaults were now reduced to levels well below 1 roentgen per hour, and the bodies were deemed safe for transport.

Eighteen days after the accident, military planes carried the vaults to airports near the intended burial grounds. McKinley's family chose Arlington National Cemetery on account of his service in Korea, as well as a sense that he had died in service to his country. Byrnes and Legg were buried in cemeteries in New York and Michigan respectively.

The funerals were bizarre affairs. On the day before each funeral, an extra-deep grave was excavated, and a foot-thick layer of concrete was poured and allowed to harden, forming a pad. On the day of the funeral, the rites were limited to eight minutes, with the grieving family members standing at least twenty feet away from the vaults. Legg's family insisted on the casket being taken out of its vault for the rites. This caused radiation levels to double, and the service was limited to five minutes. The vaults were then lowered onto the pads and, after the families had left, more concrete was poured so that the vaults were completely encased in foot-thick concrete—two feet thick, in Legg's case. Finally, the concrete was covered with several feet of earth.

In their effort to establish the cause of the accident, investigators took two main directions: interviews with SL-1 staff and administrators, and examination of the accident site. The staff interviews produced only limited information. The only eyewitnesses to the accident were dead, after all. The workers who had shut down the reactor before the Christmas holiday described how three of the control rods refused to fall freely into the reactor core and had to be driven down. This sticking behavior was much worse, they reported, than had been experienced earlier.

Engineers described another problem that had developed with the reactor over its two-year operating lifetime. As part of the reactor design, strips of boron had been attached to the fuel plates. Boron, like cadmium, is an efficient absorber of neutrons, and the strips had the function of helping prevent a runaway reaction and extending the life of the core. Nevertheless, the strips had begun to corrode and flake away from the fuel plates, leaving the reactor more excitable than it had been when it came online. Thus, the control rods didn't have to be raised as far as they did previously in order to start the nuclear chain reaction.

On account of these and other problems that had developed with the SL-1 reactor, officials had planned to replace the entire reactor sometime during 1961. When the accident occurred, therefore, the reactor was being operated on borrowed time. The crews had to keep recalibrating the control rods, for example, to ensure that the control motors moved them up and down within the range that controlled the reactor core. Nevertheless, no one believed that the reactor was threatening to get out of control. There still seemed to be a large margin of safety built into the various operations required for starting, stopping, and running the reactor.

Officials described the background and training of McKinley, Byrnes, and Legg. Because McKinley was a new trainee, there was some suspicion that he might have done something to trigger the accident. Attaching the control rods to the drives—the task that the men were engaged in at the time of the accident—required the rods to be raised by hand by about four inches. Raising the central rod by more than about sixteen inches would trigger a runaway chain reaction (a nuclear "excursion," as it is called). Both Byrnes and Legg were well versed in the procedure and knew not to lift the rods beyond the permitted four inches. Could McKinley, out of ignorance, have pulled the central rod much farther out than that?

During the first few days of the inquiry this scenario seemed especially plausible on account of an unfortunate circumstance: In the haste and confusion of the recovery, the bodies were misidentified. McKinley's body was originally thought to be the one that was impaled in the reactor room ceiling. This would have meant that he was standing directly on top of the reactor at the time of the accident, where he could have been handling the central control rod. In the course of the autopsy, however, it became clear that McKinley was the man who was pulled out first—the man who was still alive at the time the rescuers arrived. This man had been at some distance from the reactor at the time of the accident, thus he could not have been handling any control rod. Once the bodies were correctly identified, the idea that the accident was the result of a trainee's mistake fell apart.

Much more information was obtained from the reactor site. During the eleven months following the accident, workers gradually tore down the reactor building while carefully documenting every item that was found. Many of the items, including the control rods, were extremely radioactive. These were taken for study to a "hot lab" on the Testing Station, where they could be handled, cut up, and inspected with remotely controlled instruments.

The investigators first wanted to establish whether or not a nuclear excursion had in fact occurred. The fact that radiation levels in the reactor building were so high didn't compel that conclusion: The radiation came primarily from nuclear fuel that had been ejected from the reactor, but the fuel might have been ejected as the consequence of a chemical explosion or some other event within the reactor vessel. To resolve this issue, radiation technologists took one of the dead men's wedding rings and dissolved it in acid. They found that some of the gold atoms in the ring had been converted from normal gold, ^{197}Au, to radioactive gold, ^{198}Au—a transition that occurs by capture of an extra neutron. Thus the presence of ^{198}Au was proof that there had been an intense flash of neutrons during the accident, and these neutrons must have been generated by an uncontrolled chain reaction within the reactor's fuel elements.

The central control rod (minus its cadmium blade) was one of the early items to be recovered, because it was found lying directly on top of the reactor. Evidently it had fallen there after being violently ejected during the explosion and striking the reactor room ceiling. The blade had broken off and remained inside the reactor.

From the control rod, the investigators learned the exact moment in the proceedings when the accident occurred. To reattach a control rod to its drive, the crew had to take the following steps, which formed an unvarying routine. First, they had to insert a handle into the top of the rod, effectively extending the rod by a couple of feet.

Second, a crew member grasped the handle and raised the control rod by about four inches, while another worker attached a C-clamp to the rod. The rod was then lowered back down until the C-clamp rested against the housing through which the rod passed as it entered the reactor vessel. This held the rod in a fixed position for the following steps. The handle was removed and a nut and washer were attached to the top of the rod—these were required for attachment to the drive. Then the handle was reattached and a worker raised the rod just slightly—maybe a quarter of an inch—so that another worker could remove the C-clamp. The handle was removed, and the drive was connected to the top of the rod.

When it was recovered, the control rod had nothing attached to its top end. In fact, the very topmost part of the rod had been broken off and was missing. Then workers found this missing part in the attic above the reactor. Attached to this part was the washer and nut used for joining the rod to its drive, and attached to the washer and nut was the lifting handle.

The fact that the handle had been attached to the central control rod meant that the crew was working on this rod when the accident occurred. Furthermore, since the washer and nut were already in place between the handle and the rod, the workers must have already completed the first, four-inch lift and attached the C-clamp and the washer and nut. They were now due to perform the second, tiny lift that was required to release the C-clamp.

This wasn't all that was learned from the central control rod. When it was found lying on top of the reactor, the rod still lay within its "shield plug"—a kind of steel collar normally fixed to the top of the reactor vessel, through which the rod could be moved up and down. Below the shield plug was fixed a "guide tube": This formed a sheath around the rod, which moved up and down inside it. Evidently the rod, the shield plug, and the guide tube had been blown out of the reactor as a unit. When found, the rod was only slightly withdrawn from the plug, just as would be expected for the phase of the operation that was being performed at the time. This seemed to argue against the idea that someone had withdrawn the rod the sixteen inches required to initiate a chain reaction in the core.

But quite a different story emerged when the rod and the plug were disassembled and examined microscopically. The control rod was peppered with small impact marks and scratches that had been caused by the explosion. So was the guide tube. It quickly became apparent that the individual marks on the guide tube could be lined up with corresponding marks on the control rod—but only if the top of the rod was withdrawn from the shield plug by twenty inches. Thus the rod must have been raised by that amount at the time of the explosion.

Apparently, someone or something had raised the rod much farther than was prescribed or permitted, and in fact four inches beyond the distance needed to set off an uncontrolled chain reaction. Evidently, when the rod/plug assembly was propelled violently against the ceiling, the tip of the rod (along with the handle) broke off, and the remainder of the rod was driven back down through the shield plug and guide tube, leaving it in a seemingly normal position.

The investigators confirmed the excessive withdrawal a few months later when workers were able to inspect the reactor core. The cadmium blade of the central control rod was trapped by the partially melted and deformed fuel plates, but it was twenty inches above its lowermost position. Thus the entire control rod assembly had been withdrawn by that distance at the time of the accident.

Why had the control rod been raised so far beyond the safe level—especially at a phase of the operation when it needed to be raised by only a fraction of an inch? The most obvious explanation was that the rod had become stuck. A crew member—or two crew members working together—might have tugged mightily to free it. Then, when the rod suddenly broke loose, they unintentionally pulled it far beyond the intended distance.

This scenario sounds entirely plausible: Everyone has had similar experiences in the course of wrestling with balky mechanical devices, if not with such disastrous consequences. But could it really have happened at SL-1? To find out, the investigators built a replica of the control rod that could be locked and released at will. They had a large number of workers attempt to free the "stuck" rod by pulling on it as hard as they could. When the rod was suddenly and unexpectedly released during the pull, none of the workers raised the rod more than about ten inches. This was certainly an uncomfortably large distance, but it wasn't enough to start an uncontrolled chain reaction in the core. The main reason that the rod didn't come up very far when it was released was that it was so heavy—about one hundred pounds. As soon as the workers' initial effort slackened, it stopped dead or fell back.

The investigators considered another scenario, which was that one of the crew members had played a prank on the man who was tugging on the rod, perhaps by grabbing or pinching him in the rear. A reflex response might have led the man to straighten up, causing him to pull the rod upward by twenty inches. To test this idea, the investigators actually perpetrated this prank on a number of men who were in the process of pulling on the mock-up rod. None of the men who were goosed in this way pulled the rod up by any extra distance, however.

Eventually, the identity of the crew member who raised the control rod became clear: It was Jack Byrnes. McKinley, as already mentioned, was not within reach of the rods when the accident occurred. Legg must have been crouching astride control rod number seven, one of the outer control rods, because this was the rod that ripped through his body, sending him rocketing upward and pinning him to the ceiling. Because of his crouched posture, he was far too low down to have been pulling on the central rod, but he was at an ideal height to detach the C-clamp from it. Furthermore, the pattern of radiation absorbed by the bodies of Legg and Byrnes confirmed that Legg was crouching down, whereas Byrnes was standing above the center rod, in the normal position for pulling on the rod.

All in all, it seemed clear that the accident happened because Jack Byrnes, an experienced operator, raised the central control rod too far, but the exact reason for his action remained obscure, at least in the official report. The investigators wrote that "the reason or motive for the abnormal withdrawal is considered highly speculative, and it does not appear at all likely that there will ever be any reason to change this judgment."

Calling something "highly speculative" is of course an invitation to speculation, and there has been plenty of that in the years since the SL-1 accident. For the most part, it has revolved around an unproven theory that Jack Byrnes perpetrated the explosion as a murder/suicide—his motive being to bring to an end a desperate love-triangle involving himself, his wife, and his fellow crew member Dick Legg.

Even while the original inquiry was going on and radiation specialists were analyzing the remains of the reactor, investigators were also checking into the backgrounds and family circumstances of the three dead men. None of their findings made it into the report, but they soon became the subject of gossip around the Testing Station. In July 1962 Leo Miazga, an investigator for the Atomic Energy Commission (AEC), filed a supplemental report describing what he had learned about the crew members and their families. Miazga's report was never made public, but recently a copy was obtained by Colorado-based journalist William McKeown, who wrote a book about the SL-1 accident.

According to McKeown, Miazga's report documented the breakdown of Jack and Arlene Byrnes's marriage. Jack had been spending less and less time at home. On the evening of the accident he had taken his personal items with him in his car as if he was moving out. When Arlene called Jack at work later that evening, it was to discuss ending their marriage. Thus Jack Byrnes may have been in a distracted mood—or worse—during the later part of the work shift.

Miazga did not find evidence to link Dick Legg to Arlene. Still, it was clear that Byrnes and Legg did not get along. In fact, they had engaged in a drunken fistfight during a bachelor party in May 1960, eight months before the accident.

The first that the wider public heard about a possible love triangle as the cause of the accident was in 1979, when a Vermont newspaper, the *Brattleboro Reformer*, ran a story about an AEC memo about nuclear safety. The memo had been written in 1971 by Stephen Hanauer, then an AEC staff member. Hanauer had cited the love-triangle story to illustrate the notion that nuclear-reactor accidents could be caused by internal sabotage. Hanauer's memo was leaked to the *Reformer* by Robert Pollard, a nuclear regulator who became an antinuclear campaigner. The *Reformer's* story was picked up by the national press, and two years later a television documentary about the accident mentioned the love-triangle theory. McKeown's book gives it a great deal of play without drawing any firm conclusions. When I spoke with Stephen Hanauer, now retired, in 2005, he portrayed the love triangle as a mere piece of gossip that he had picked up, not something that was backed by any kind of solid evidence.

There always will be the temptation to attribute accidents to the foibles of individuals—especially deceased individuals—because doing so may be seen as relieving planners, designers, and administrators from responsibility. A case somewhat similar to the SL-1 accident occurred in 1989, when an explosion destroyed a gun turret on the U.S. battleship *Iowa* during training exercises, killing forty-seven sailors. The Navy initially called the explosion a suicidal act by one of the gunners, Clayton Hartwig. Hartwig was said to have been depressed on account of a sexual relationship with another sailor that had gone sour. Later, the evidence for this explanation fell apart, and the Navy apologized to Hartwig's family. The explosion was probably caused by accidental over-ramming of the gun's propellant charges, perhaps combined with faulty packing of the charges.

Whatever the reason why Jack Byrnes raised SL-1's central control rod so far, it was quickly recognized that allowing the movement of a single control rod to trigger an uncontrolled nuclear chain reaction was a serious design flaw. This flaw came about primarily on account of the small size of the SL-1 reactor, which used a total of only five control rods. Every subsequent reactor design has required the lifting of several control rods to cause a runaway reaction. Still, even with such reactors it is possible to trigger nuclear excursions by inappropriate lifting of multiple rods. In fact, an accident of that kind happened in Canada, nine years before the SL-1 accident. During a crisis triggered by other events, an operator at the Chalk River reactor in Ontario erroneously pressed a button that raised several banks of control rods. As a result, a nuclear excursion began that blew the four-ton lid of the reactor vessel into the roof of the building. The accident contaminated the entire plant, but no one was killed.

The worst reactor accident in the United States, the partial core meltdown at the Three Mile Island reactor in Pennsylvania in 1979, was caused by loss of coolant to the core, not by problems with the control rods. There was only an insignificant release of radioactivity to the environment, and no one was killed or injured. The world's worst nuclear power plant accident—and the only accident at a commercial plant that has ever resulted in radiation-induced deaths—was the 1989 disaster at the Chernobyl-4 reactor in Ukraine. This was caused by operator errors during a test of the reactor's response to a loss of electrical power supply, combined with design deficiencies. More than fifty people—mostly emergency workers—died of radiation injuries in the aftermath of the accident. Some experts believe that larger numbers of people in the general population are dying or will die from radiation-induced illnesses.

The SL-1 accident may have had other victims. Hazel Leisen, the nurse who tended to Richard McKinley while he was in the ambulance, died of cancer a few years after the accident. Ed Vallario, the health physicist who led the rescue attempt, and two of his helpers, also died of cancer, although Vallario didn't fall ill until more than thirty years after the accident. The statistics of small numbers don't permit any firm conclusion as to whether or not the four cancer deaths resulted from the radiation exposure these individuals received in 1961.

CHAPTER 8

⚮

MICROBIOLOGY:
Gone with the Wind

On April 4, 1979, Anna Komina began to feel unwell. Her symptoms were not very specific: a feeling of faintness and dizziness, combined with a shortness of breath. After a day or so, she felt better, but on April 8 she collapsed. A doctor was called. He found that her blood pressure was dangerously low, and he summoned an ambulance. The emergency medical technicians struggled for several hours to raise her blood pressure to the point that she could be safely transported to the hospital. Two days later, she died.

Komina, who was fifty-four years old at the time of her death, was a resident of the city of Sverdlovsk, 850 miles east of Moscow, on the far side of the Ural Mountains. (The city is now known by its prerevolutionary name of Ekaterinburg.) She lived in a modest cottage in an industrial southern district of the city known as Chkalovskiy, along with her husband, son, daughter-in-law, grandson, and two pigs. She worked at a nearby ceramics factory, where her job was to check the temperature and pH of the water supply.

After Komina died, her body was autopsied. Her death was officially attributed to bacterial pneumonia. Then the body was doused in chlorinated lime, placed in a steel coffin, loaded onto a truck, taken to a special section of a local cemetery, and buried—all at state expense. Komina's husband was not invited or even informed of the burial. He found out anyway and went to the cemetery, but the police would not let him in. Not long afterward, he died of grief.

Komina was not the only citizen of Sverdlovsk to die unexpectedly during the spring of 1979. During the six-week period starting on April 6, at least seventy-six other women and men came down with symptoms similar to hers. Massive doses of antibiotics and other medicines saved eleven of them, but at least sixty-six persons died. Some died so quickly that they did not make it to the hospital alive. What killed them was anthrax.

The first that anyone in the West heard about the Sverdlovsk anthrax outbreak was in October of 1979, when a Russian-language newspaper catering to the émigré community in Germany ran a brief story alleging that thousands of Russians had died of the disease. In the same month the U.S. Central Intelligence Agency received descriptions of the outbreak from recent Russian émigrés. The mention of Sverdlovsk as the site of the outbreak raised a red flag, because the CIA suspected that a Soviet bacteriological institute in that city was engaged in germ-warfare research. The institute was housed in a high-security area known as Compound 19 in the northern part of Chkalovskiy district. Nevertheless, the CIA analysts initially discounted the idea that the release of a disease agent from that facility might have caused the outbreak. "The probability is low that the Soviets were working with a quantity of highly lethal pathogens sufficient to cause 40 or more deaths without possessing either an effective vaccine or antidote," the analysts wrote in a top-secret report.

The Sverdlovsk outbreak gained much wider publicity early the following year, when Germany's mass-market tabloid, the *Bild Zeitung*, ran two reports about it. *Bild* never missed an opportunity to badmouth the Soviets, and it milked this story for all it was worth. The outbreak was caused by the explosion of anthrax-containing munitions at a germ-warfare institute, the newspaper announced, with subsequent release of the deadly bacterium into the atmosphere. Hundreds of people were affected: Within seconds of inhaling the organisms they developed breathing difficulties, and death followed in a few hours. The CIA, *Bild* reported, had confirmed that a germ-warfare institute was the source of the anthrax release.

Bild is not known for scrupulous fact-checking, and many details of its account were clearly wrong. Nevertheless, it was true that by early 1980 the CIA had become more suspicious that the anthrax outbreak originated in Compound 19. For one thing, the CIA learned that Soviet Defense Minister Dmitri Ustinov had made a visit to Sverdlovsk at the time of the outbreak, suggesting some involvement of the military. Also, when CIA analysts went back and scrutinized photographs of Sverdlovsk taken by spy satellites at the time of the outbreak, they noticed some odd things such as roadblocks and what looked like decontamination trucks. Also, roads that had previously been dirt had suddenly acquired a freshly tarred surface. These might all be signs of an organized response to an epidemic caused by airborne pathogens.

In March of 1980 the U.S. State Department made their suspicions public. Their statement, though couched in terms of "indications" rather than factual allegations, amounted to a serious indictment of the Soviet Union, because that nation, along with the United States and most other countries, had signed a Biological and Toxin Weapons Convention. This convention, which went into force in 1975, banned the development and production of bacteriological weapons. The treaty didn't prohibit the Russians from doing research into *defenses* against biological agents, such as vaccines, but the airborne release of enough anthrax to kill people over a wide area suggested that the bacteria were being produced in quantities far beyond what was necessary for vaccine development.

As might be expected, the Soviet response was one of indignation and injured innocence. Yes, there was an outbreak of anthrax, they said, but it was an entirely natural one. The outbreak had started among farm animals, and it had spread to humans when meat from the slaughtered animals had been sold on the black market and eaten.

The anthrax organism, *Bacillus anthracis,* is present in many soils in the form of inert spores. It's not unusual for grazing animals to become infected by ingestion of these spores—outbreaks or sporadic cases of gastrointestinal anthrax occur among farm animals in Russia from time to time. And humans can indeed develop gastrointestinal anthrax from eating infected meat. They can also develop skin lesions (cutaneous anthrax) from handling infected animal products such as hides.

Still, a couple of facts made the Soviet explanation a bit difficult to accept. For one thing, farm animals would not yet have been put out to pasture at the beginning of April—Sverdlovsk is in Siberia, after all—so their opportunities to acquire an infection were limited. Also, some of the symptoms reported among the human victims, such as shortness of breath, were more suggestive of inhalation anthrax—a rarer and more dangerous form of the disease in which the anthrax organisms enter the body through the thousands of tiny air sacs in the lungs.

Eventually, the CIA analysts settled on a kind of compromise hypothesis. The large cluster of human cases in early April must indeed have been caused by an airborne release of anthrax spores, they concluded, but the "tail" of cases that occurred over the following six weeks probably had some other explanation. That was because, according to most medical authorities, the incubation period for anthrax is very short, so if a person doesn't fall ill within two or three days of exposure, he or she will never do so. Perhaps these late cases were indeed gastrointestinal anthrax, caused by consumption of meat from animals that had themselves inhaled anthrax during the initial exposure event.

Lacking sufficient expertise in the field, the CIA consulted with a number of American scientists about the Sverdlovsk episode, including the Harvard molecular biologist Matthew Meselson. Meselson was one of the scientists who made notable contributions during the golden age of molecular biology in the 1950s and 1960s. He is most famous for a set of experiments, conducted with Frank Stahl, which proved that the double helix of DNA copied itself in the fashion predicted by Francis Crick and Jim Watson—by separating its two strands, each of which then provided the template for a new partner. He is the kind of scientist whom everyone assumes has won a Nobel Prize, but actually he hasn't—not yet, at least. Now in his mid-seventies, he still runs an active research program, which is now focused on understanding why evolution favors sexual over nonsexual reproduction.

Since the 1960s, Meselson had been a strong proponent of restrictions on chemical and biological weapons, and he had worked from time to time as a consultant to the U.S. government on the issue. He considered the international convention banning bioweapons manufacture to be an important milestone, and perhaps for that reason he was not particularly inclined to believe that the Soviets were violating their treaty obligations. Whatever his motivation, he expressed some skepticism about the CIA's allegations.

Another series of events during the 1980s may have served to further undermine the CIA's credibility in Meselson's mind. This was the famous "yellow rain" controversy. In 1981 U.S. Secretary of State Alexander Haig publicly accused the Soviet Union of providing highly poisonous chemicals derived from fungi (mycotoxins) to its Communist allies in Vietnam and Laos, who had then used them in attacks against rebellious Hmong tribespeople. The Hmong had described the toxins as falling out of the sky in the form of a yellow rain, and said that persons touched by it quickly fell ill or died. The Hmong provided the CIA with samples of foliage covered by spots of "yellow rain," and analysis of these spots supposedly revealed that they contained mycotoxins. These findings were the basis of Haig's allegations.

Very soon thereafter, British scientists discovered that the spots contained partially digested pollen. This led Matt Meselson to hypothesize that the yellow spots were nothing more than bee feces. With other scientists, Meselson went to Thailand and actually observed a shower of "yellow rain" as a swarm of Asian bees engaged in a mass cleansing flight. He even found "yellow rain" on the windshields of cars in Harvard University's parking lots. Further analysis of both the original samples and newly obtained ones failed to confirm the presence of mycotoxins. While it is still a possibility that chemical agents were used against the Hmong, the yellow rain itself seems to have been an entirely natural and harmless phenomenon.

The yellow-rain experience may have helped influence Meselson's attitude about the Sverdlovsk allegations. Indeed, some recent accounts portray Meselson as becoming a flat-out nonbeliever in the

CIA's germ-warfare theory during the course of the 1980s. Judith Miller, Stephen Engelberg, and William Broad—all writers for the *New York Times*—cited several statements by Meselson in their 2001 book, *Germs,* that seemed to indicate his acceptance of the Soviet explanation. In 1989, for example, they quote Meselson as testifying as follows before a Senate committee hearing: "The burden of the evidence available is that the anthrax outbreak was the result of a failure to keep anthrax-infected animals off the civilian meat market, as the Soviets had maintained, and not the result of an explosion at a biological weapons factory as previously asserted by the United States." And he added a general endorsement of the success of the Biological Weapons Convention. "Today, to the best of my knowledge, no nation possesses a stockpile of biological or toxin weapons."

When I interviewed Meselson in 2006, he expressed some irritation with this portrayal. "I didn't say that [the Soviet explanation of the Sverdlovsk episode] was necessarily correct," he told me. "They should have quoted what was on either side of that, which was that we really didn't know, and they only way to find out would be to go there and find out by an independent investigation." As Meselson paints it, his attitude at the time was one of healthy scientific skepticism toward either side's account. "There are a number of reasons why you might repave a road. There might be many reasons why a minister of defense would go to a certain city—he might have a girlfriend there, even. For me, coming from peer-reviewed science, we have to look more carefully: Let's create some hypotheses and see if we can disprove them."

During the 1980s, Meselson did in fact make several attempts to organize a fact-finding trip to Sverdlovsk, but he was repeatedly stymied by a lack of cooperation on the part of Soviet authorities. This was probably due in part to their reluctance to allow anyone to investigate the anthrax outbreak, but in addition Sverdlovsk was generally off-limits to foreigners because it was a key node in the Soviet military-industrial complex.

Meselson did go to Moscow in 1986, however, where a high-level Health Ministry official by the name of Pyotr Burgasov told him that the anthrax outbreak had been caused by contaminated cattle feed.

This story explained how an animal outbreak could have occurred before animals were put out to pasture.

Two years later Meselson brought Burgasov and two other officials to the United States, where, in a series of lectures, they enlarged on this story. Most of the victims, they said, died of intestinal anthrax—and some of cutaneous anthrax that developed into a systemic infection—but none had inhalation anthrax, the form of the disease that would most likely have resulted from the release of anthrax spores into the atmosphere. Meselson seemed to accept this account, because he made public statements supporting its plausibility and once again criticized the official CIA viewpoint.

With the new policy of openness promulgated by Mikhail Gorbachev in 1990, stories began appearing in the Soviet press that linked the Sverdlovsk outbreak not to tainted meat, but to activities at the military microbiological institute. Later that year, the *Wall Street Journal* published a series of three articles on the outbreak by Peter Gumbel, then its Moscow bureau chief. Gumbel had traveled to Sverdlovsk and interviewed some of the relatives of the victims. Tamara Markova, for example, recounted how her husband Mikhail, then forty-six, had come down with a cough during the first week of April, 1979, and died shortly afterward. "The doctor said his lungs looked like jellied meat," Tamara told Gumbel. This and other accounts, as well as interviews with persons who were familiar with the autopsies that had been performed on the victims, persuaded Gumbel that the anthrax infections had been acquired by inhalation, not by consumption of tainted meat.

Meanwhile, inside information about the alleged Soviet germ-warfare program became available. In late 1989 Vladimir Pasechnik, the director of the Institute of Ultrapure Biological Preparations in Leningrad, defected to Britain, where he recounted to intelligence agents how his institute had been working on methods of delivering bacteriological agents by means of cruise missiles. His own institute, Pasechnik said, was just one element of a massive network of germ-warfare facilities known as Biopreparat, which employed 30,000 workers across the Soviet Union. Lethal bacteria such as plague and anthrax were manufactured and stored at these facilities, ready to be loaded into munitions and delivered by specially adapted planes, cruise missiles, and even intercontinental ballistic missiles. The weapons had been successfully tested on tethered monkeys on an island in the Aral Sea. The Soviet Union, if Pasechnik's statements were to be believed, was violating the Biological Weapons Convention in the most flagrant way, though ironically Pasechnik said that he had never heard of the convention. Pasechnik's allegations did not become public knowledge until 1992, and he apparently did not provide specific information about the cause of the Sverdlovsk anthrax outbreak.

After the breakup of the Soviet Union in 1991, many embarrassing secrets of the Soviet era were exposed. In November 1991, a Russian general told the newspaper *Izvestiya* that the Sverdlovsk anthrax outbreak had originated at the bacteriological institute when workers had failed to activate safety filters, leading to a massive release of the bacterial spores into the atmosphere. In May of the following year Russian President Boris Yeltsin (who had been the Communist Party boss in Sverdlovsk at the time of the anthrax outbreak) officially admitted that the outbreak had been caused by the Soviet military. He said that he gave orders for the bacteriological-warfare program to be terminated.

"In a sense you could say that he scooped us," Matt Meselson said of Yeltsin's announcement. Scientists, like journalists, hate to be scooped: Priority is everything. One way that scientists often deal with this problem is by denigrating the significance of other people's findings that anticipate their own. In this case Meselson suggested to me that Yeltsin didn't really have any solid reason to make the statement he did. "It was pretty slim evidence," he told me. It wasn't based on an actual admission of guilt by the military but on an unverified statement from one of Yeltsin's advisors—a biologist named Alexey Yablokov—to the effect that "someone had found anthrax spores on a wall hanging of some kind."

Peter Gumbel, the *Wall Street Journal* reporter, received the same treatment. Gumbel's account was so full of errors as to suggest that he never visited the area affected by the outbreak, according to a 2002 article by Boston College sociologist Jeanne Guillemin, who is Meselson's wife and one of his collaborators in the Sverdlovsk study.

In a recent e-mail to me, Gumbel fired back. "I had just driven a 10-ton truck through [Meselson's] credibility," he wrote. "He had staked his academic reputation on some Cold War lies. At best, he had been duped by some fairly crude propaganda. At worst, he was a naive apologist for a nasty regime. I imagine his Harvard colleagues snickered behind his back. . . . And then his wife comes back and tries to trash me in her official account. I was surprised by the pettiness. It would have been more professional to acknowledge the groundwork that I did. . . . As far as I know, I didn't get my facts wrong. Unlike certain others. . . ."

In any event, the gradual flow of revelations made Meselson's willingness to believe the Soviet denials less and less tenable, but they also gave him the opportunity he had long wished for—to visit Sverdlovsk and find out the truth for himself. In early 1992, with Yablok-ov's help, Meselson obtained an invitation from Ural State University to come to Sverdlovsk to carry out an investigation.

In June of 1992 a team of five researchers went to Russia: Matt Meselson, Jeanne Guillemin, Alexis Shelokov, David Walker, and Martin Hugh-Jones. Guillemin spoke some Russian, and her role would be to carry out and analyze interviews with families and survivors. Shelokov was a public-health expert from the Salk Institute and a native Russian speaker. Walker, a pathologist from the University of Texas, would review any autopsy data they could find. Hugh-Jones, a veterinarian from Louisiana State University, would be in charge of investigating the anthrax outbreak as it affected animals. The team gave themselves two weeks to unravel a thirteen-year-old mystery.

In Moscow, the group met with Pyotr Burgasov, the Health Ministry official (now retired) who had presented the tainted-meat story to the world. He stuck by this story, even though it was by now widely discredited and he himself was being portrayed in the Russian media as an accomplice of the KGB. Then the group traveled on to Sverdlovsk.

Meselson and his colleagues wanted to find out who had contracted anthrax, where and when they had contracted it, what their symptoms were, and what the autopsies revealed. Unfortunately, most of the relevant documents such as hospital case records were missing:

They had been confiscated by the KGB soon after the 1979 outbreak. Meselson did know, however, that the anthrax victims were buried in a special section of a local cemetery. Therefore, on Saturday, June 6, the group went to the cemetery, where they indeed found a section containing sixty-six graves of men and women who had died in the six-week period beginning April 8, 1979. These included the graves of Anna Komina, whose death was recounted at the beginning of this chapter, and Mikhail Markov, the man whose widow was interviewed by Peter Gumbel. Interestingly, there were no children among the victims. In fact, there were no family groupings: The victims were unrelated to one another. This was somewhat inconsistent with the tainted-meat story, because one might expect several family members to be struck down if tainted meat was served.

The researchers now had a list of the victims' names, ages, and dates of death, but not yet their places of residence or work, nor any means to contact their relatives. But more information, including some of the victims' home addresses, was provided by Margarita Ilyenko, a physician who was the director of one of the hospitals that received the anthrax victims. Ilyenko had not only treated the victims and saved some of their lives; she had also helped organize the community response, recruiting volunteers to survey the affected households, carry out disinfections, and the like. She described how the bodies of the victims had overflowed the hospital's morgue and piled up outside, while no one dared go near them for fear of catching whatever had caused their death.

Later, toward the end of the team's visit to Sverdlovsk, a Supreme Soviet deputy by the name of Larissa Mishustina handed the team an official list of sixty-four victims with their addresses. The KGB had prepared the list after Boris Yeltsin ordered the victims' families to be compensated—something that was never actually done.

With this information, Jeanne Guillemin began touring the streets of the Chkalovskiy district, where many of the victims had lived. As an interpreter she had Olga Yampolskaya, a Moscow physician who had joined their team. During the anthrax outbreak Yampolskaya had been an assistant to the head clinician dealing with the outbreak, and she had herself treated some of the victims. Guillemin's and

Yampolskaya's task was to knock at the addresses on her list and ask if the current resident was a relative of the deceased. In the Soviet era people rarely moved, and although thirteen years had passed, a surprisingly large number of the homes she visited were still occupied by the victims' spouses or children.

When they knocked at the address given for Fagim Dayanov on Escadronaya Street, for example, they found his widow, Rema. She told them that Fagim, like several other victims, had been employed at the ceramics factory, which was situated three kilometers from Compound 19 in a southeasterly direction. Fagim had remained healthy during the first few weeks of the outbreak. When the cause was identified as anthrax, vaccinations were made available; Rema agreed to be vaccinated but Fagim refused. On May 3, probably as part of the disinfection campaign, Fagim was told to clean the roof of the factory. He became ill on May 4 and was taken to the hospital on May 6. Rema and their son were not allowed to visit him until three days later, by which time he was desperately ill. The couple was kept separate by a glass partition and had to communicate by exchanging notes. On the evening of May 10 Fagim wrote: *Rema—don't go home. We may never look into each other's eyes again.* Rema eventually did go home, and Fagim died that night. When Rema returned to the hospital, not knowing that her husband was dead, she was directed to the morgue, where her husband's body lay on a gurney, naked, with his chest, abdomen, and braincase sliced open. The autopsy had already been completed.

Besides these often-painful interviews with the victims' relatives, Meselson's team had extensive discussions with the two pathologists, Faina Abramova and Lev Grinberg, who had conducted most of the autopsies. It was Abramova who made the initial diagnosis of anthrax, based on the observation of the anthrax bacillus in tissue samples. She related how the KGB had come to her hospital to confiscate all the medical and autopsy records and tissue samples, but she successfully hid much of the material: She placed jars containing tissue samples among other items in her pathology museum, and she concealed paperwork in filing cabinets devoted to other matters. Thus Abramova and Grinberg were able to give Meselson's team a comprehensive description of forty-two cases. All of them had shown signs of infection in the thoracic cavity, with bleeding from thoracic lymph nodes and from the mediastinum (the area between the left and right lungs). Some victims also had involvement of the gut, but this involvement was probably secondary to a systemic infection, because the mesenteric lymph nodes (the nodes that receive lymph from the intestines) were affected in only a few cases. In short, the pathological findings were consistent with inhalation anthrax in most or all cases, just as Peter Gumbel had concluded in his *Wall Street Journal* articles the previous year.

The two weeks that Meselson's team had allotted themselves for the trip to Russia came to an end before they were able to carry out most of their interviews with the victims' relatives, and they resolved to make a return trip the following year. Nevertheless, even with the results that they obtained on the 1992 visit, two things were clear: The outbreak was caused by the airborne release of anthrax spores and, second, most of the victims either lived or worked in the Chkalovskiy district of Sverdlovsk at the time they became ill. The researchers' findings were thus pointing strongly toward the conclusion that Meselson had long resisted, namely that the outbreak was caused by an accident at Compound 19, the suspected germ-warfare institute.

In the fall of 1992, an event occurred that to some degree made Meselson's investigations irrelevant. The second-in-command at Biopreparat, the network of Soviet facilities that Vladimir Pasechnik had fingered as a germ-warfare agency, defected to the United States. His name was Kanatjan Alibekov, but he later anglicized it to Ken Alibek. When interviewed by the CIA, Alibek poured out a truly horrifying account of the Soviet germ-warfare program to his CIA handlers, a description that went far beyond what Pasechnik had revealed. Among other things, he said that the Soviets had produced hundreds of tons of anthrax. The lethal dose of anthrax when inhaled is about ten thousand spores or ten billionths of a gram, so the Soviets had manufactured enough of the spores—in theory—to kill the entire population of the planet millions of times over. Alibek identified Compound 19 as being heavily involved in research,

production, and weaponization, and he said that he himself had developed a strain of anthrax that was far more lethal than those that
occurred in nature.

Meselson was still determined to complete his own study, and
in August of 1993 he and Guillemin returned to Sverdlovsk. They
interviewed more relatives of the deceased victims, as well as some
of the few victims who had recovered, and they obtained detailed
information about the victims' whereabouts during the outbreak.
In addition, Meselson and Guillemin visited several villages outside
of Sverdlovsk where farm animals had come down with anthrax
during the same period. One place they were not allowed to visit,
however, was Compound 19, the suspected site of anthrax release.
Although anthrax production had probably long ceased, it was still
a military facility, and Meselson could not obtain the required au-
thorization to enter.

Back in the United States, Meselson and Guillemin put all the
information together in an attempt to pin down the time and place of
the anthrax release. The first person to fall ill, a forty-eight-year-old
ceramics factory employee named Vera Kozlova, did so on Wednesday, April 4, so there must have been some release of anthrax before
then—probably a day or two before, given what was known about
the minimum incubation period for inhalation anthrax. So many victims fell ill in the few days after Kozlova was struck down that it
seemed likely that a single release event very early in that week had
caused most of the infections.

Looking at their records, Meselson and Guillemin saw that nearly
all the victims either lived in the Chkalovskiy district to the southeast
of Compound 19, or they worked in or visited that district in the relevant time span. Six victims lived or worked in Compound 19 itself.
As already mentioned, several of the victims worked at the ceramics
factory southeast of Compound 19. Another group of victims were
five military reservists. These men all lived and worked outside of
Chkalovskiy, but during the week starting Monday, April 2, all five
took a course at Compound 32, a military facility immediately to the
south of Compound 19. The reservists came in each morning and left
the district in the late afternoon. This strongly suggested that the an-

thrax release occurred during daytime hours on one of the weekdays from April 2 to April 6, most likely on April 2 or April 3, given that the earliest disease onset was on April 4.

When the researchers took a map of Sverdlovsk and plotted the daytime locations of all the victims during this time span, a striking pattern emerged. Nearly all the victims were located in an extremely narrow cigar-shaped zone that began in Compound 19 and extended in a southeasterly direction. The closest victims—six of them—had been within Compound 19 itself. The most distant had been four kilometers away on the southeastern outskirts of the city. Although no humans were affected beyond that distance, animals were: The veterinary cases occurred in six villages that lay precisely along a continuation of that southeasterly line. The most distant of these villages lay more than fifty kilometers from Compound 19.

Five victims were "outliers"—they neither lived nor worked in the high-risk zone. Three of them had occupations, such as truck driver, that might have taken them into the zone. Information about the other cases was lacking.

Thus it was clear to the researchers that the anthrax cloud had originated at Compound 19 and the wind had blown it toward the southeast. This was the final clue that allowed Meselson to pin down the date of the accident. Sverdlovsk airport, like all major airports around the world, reports weather conditions every three hours to a United Nations agency, and this information is archived by the U.S. National Center for Atmospheric Research in Boulder, Colorado. When he obtained the records for the week in question, Meselson found that the wind had blown steadily to the southeast or south-southeast throughout the daylight hours on Monday, but never blew in that direction during the following two days. The wind did blow to the southeast for part of the day on Sunday, April 1, but this day was ruled out because the military reservists were not in the area until Monday. Thus Monday, April 2, was pinned down with near certainty as the date of the accident. As icing on the analytical cake, it turned out that one of the reservists had attended classes at Compound 19 on only a single day—Monday.

Meselson now turned to another question: How much anthrax

was released? To answer this question, he had to make a lot of "guesstimates" about factors such as the height of the release aboveground, atmospheric conditions, and the number of spores that would be required to cause a fatal infection. In the end he concluded that the total release was quite small—less than a gram (four hundredths of an ounce), and perhaps as little as a few milligrams.

The researchers published their findings in two papers. A 1993 paper in the *Proceedings of the National Academy of Sciences* presented the evidence from the pathological investigations, which pointed to the conclusion that most of the anthrax infections were acquired by inhalation and not by food consumption. In the following year Meselson's group published a paper in *Science* in which they presented the findings of their epidemiological studies, which included the time, place, and estimated amount of anthrax release.

Together, the two studies presented far more detailed and convincing evidence that the anthrax outbreak originated in the biological-warfare facility than had any previous investigation. No one reading the *Science* paper could fail to be impressed how weeks of plodding epidemiological footwork had allowed the researchers to plot a cigar-shaped "arrow" that pointed unmistakably and accusingly at Compound 19.

Still, the study did have its detractors. For one thing, there were some who felt the whole topic was moot, given that the Russians had already admitted that their germ-warfare institute caused the outbreak. Also, Meselson had not been able to answer the question that people were now asking, which was what exactly went wrong in Compound 19 that led to the anthrax release. Finally, there were some critics, such as germ-dispersal expert Bill Patrick of Fort Detrick, who felt that Meselson's estimate of the amount of anthrax released was far too low. "We hooted [when we heard his numbers]," is what Patrick told the authors of *Germs*.

The significance of the dispute over the amount of anthrax released is this: If the amount was very small, it was conceivable that it happened, not as a result of weapons production, but in the course of legitimate research or vaccine development—activities that were permitted under the Biological and Toxin Weapons Convention. Tom

Mangold and Jeff Goldberg, in their 1999 book *Plague Wars,* suggested that Meselson gravitated to a low estimate because he was clinging to his long-held hope or belief that the Soviets were abiding by the convention. They dug up some statements made by Meselson in the mid-nineties that were consistent this interpretation. Still, Meselson sticks by his estimate, and it has recently been confirmed in a reanalysis by physicist Dean Wilkening of Stanford University.

One still-unresolved mystery has to do with the "tail" of anthrax cases that extended for six weeks beyond the initial outbreak. Most of these victims clearly had inhalation anthrax, not the gastrointestinal form, so it doesn't seem likely that they got sick from eating tainted meat, as the CIA analysts had originally proposed. How then had they acquired their infections?

One theoretical possibility is that the anthrax release was not a single event but continued for several weeks, albeit at a lower rate. If this were the explanation, however, one would expect the later victims to be located in other places than in the cigar-shaped zone where the early victims lived or worked, because the wind blew in other directions on later days. In reality, they were located in exactly the same high-risk zone as the early victims.

A second possibility would be that the later victims acquired their infections from anthrax that had been deposited on surfaces during the initial release and then reentered the atmosphere at some later time. Such an explanation seems particularly appropriate for victims like Fagim Dayanov who, as described earlier, fell ill in early May, a day after he was sent to clean the roof of the ceramics factory. Did he stir up a secondary aerosol of anthrax during that cleaning operation and thus breathe in enough spores to develop an infection?

When I asked Matt Meselson about this, he acknowledged that secondary infections were a possibility, but he considered this explanation unlikely. Anthrax spores are so tiny, he told me, that once they bind to surfaces they are held there by surprisingly strong electrostatic forces and are very difficult to pull back into the atmosphere. Larger clumps of spores can be reaerosolized, but such clumps are not effective infectious agents because they do not penetrate deep

into the air sacs of the lungs, which they need to reach order to trigger an infection.

Meselson favors a third explanation for the late cases, which is
that those victims inhaled the anthrax spores on Monday, April 2,
just like the early victims, but simply took longer to fall ill—up to six
weeks in some cases. This contradicts conventional medical wisdom,
which says that the incubation period for inhalation anthrax is just
a very few days—just one day in some cases. Meselson dug up some
old studies, however, in which monkeys were exposed to anthrax by
inhalation. Some of these monkeys took weeks to fall ill. Such long
incubation periods were not observed in more recent studies, but according to Meselson that was because researchers simply didn't wait
that long—the animals were sacrificed after a few days whether they
were sick or well.

In theory, analysis of the late victims' location during the outbreak
could resolve this issue: If victims such as Dayanov had been away
from Chkalovskiy during the first week of April and then returned to
the high-risk zone before becoming ill, that would suggest that they
acquired their infection from secondary aerosols. Such cases were
not found, but the numbers are too small to make a definitive judgment on this basis. Thus, the question of how the victims in the tail
of the outbreak acquired their infection remains unresolved.

How exactly did the anthrax release come about? In 1999 Ken Alibek, the defector who ran the Soviet germ-warfare program, came
out with his own book (written with Stephen Handelman), titled
Biohazard. In it, Alibek presented a detailed account of what he was
told happened:

> On the last Friday of March 1979, a technician in
> the anthrax drying plant at Compound 19, the biologi-
> cal arms production facility in Sverdlovsk, scribbled a
> quick note for his supervisor before going home. "Filter
> clogged so I've removed it. Replacement necessary," the
> note said.
>
> Compound 19 was the Fifteenth Directorate's busiest

production plant. Three shifts operated around the clock, manufacturing a dry anthrax weapon for the Soviet arsenal. It was stressful and dangerous work. The fermented anthrax cultures had to be separated from their liquid base and dried before they could be ground up into a fine powder for use in aerosol form, and there were always spores floating in the air. Workers were given regular vaccinations, but the large filters clamped over the exhaust pipes were all that stood between the anthrax dust and the outside world.

After each shift, the big drying machines were shut down briefly for maintenance checks. A clogged air filter was not an unusual occurrence, but it had to be replaced immediately.

Lieutenant Colonel Nikolai Chernyshov, supervisor of the afternoon shift that day, was in as much of a hurry to get home as his workers. Under the army's rules, he should have recorded the information about the defective filter in the logbook for the next shift, but perhaps the importance of the technician's note didn't register in his mind, or perhaps he was simply overtired.

When the night shift manager came on duty, he scanned the logbook. Finding nothing unusual, he gave the command to start the machines up again. A fine dust containing anthrax spores and chemical additives swept through the exhaust pipes into the night air.

In attributing the accident to a failure to replace a filter, Alibek's account meshed well with the statement made by a Russian general to *Izvestiya* in 1991, as described earlier. On the other hand, there are details that are inconsistent with the findings of the Meselson group, most notably the date and time of the anthrax release. If the Meselson group is right, Friday March 30 is definitively ruled out, because the wind was blowing in the exact opposite direction—toward the northwest—on that day, which would have carried the lethal plume toward downtown Sverdlovsk and away from the

Chkaloviskiy district. Also, the military reservists were not in the area at any time on that day.

In a 2006 interview, I asked Abilek how he came by the information about the accident that he presented in his book. He said that it was told to him by a senior scientist by the name of Mikhail Kuzmitsch, during a two-hour train ride in 1983. Kuzmitsch was working in the anthrax plant at the time of the accident. When I pressed him about the inconsistencies with Meselson's account, he replied with the kind of non-answer that surely reflected decades of training in the Soviet bureaucracy. "I respect his work," he said, "and he respects mine."

I suggested a possible resolution, which was that the removal of the filter did happen at the end of the day shift on Friday as Kuzmitsch told him, but that the plant lay idle over the weekend and wasn't started up until Monday, so that the release was delayed until that day. Alibek agreed that this was a possibility, and added that he wasn't sure whether the plant would have been running through the weekend.

In her book, Jeanne Guillemin gave Alibek the same treatment she gave Peter Gumbel: She took him to task for the apparent errors in his account, and suggested that they threw his entire explanation for what had happened into doubt. Meselson took a mellower view of the matter. He reminded me that details such as dates can easily change as they are stored in memory or reported to others. "The fundamentals are correct," he said. "The only thing we disagree about is the day of the week—a pretty small detail."

Given that the origin of the anthrax outbreak in the germ-warfare institute has been admitted by the president of Russia, described in detail by the onetime boss of the Soviet germ-warfare program, and proven by Meselson's research, one might think that there would be universal acceptance of this explanation. Jeanne Guillemin told me, however, that a book on infectious diseases distributed recently by the Russian Ministry of Health attributes the outbreak to—yes—tainted meat.

As far as is known, Russia and the United States now adhere to the Biological and Toxin Weapons Convention and no longer have any military stocks of anthrax or other pathogens. The country whose continued engagement in production of biological weapons has aroused the greatest concern is Iraq. Although Iraq signed the convention in 1972, Saddam Hussein instituted a biological weapons program soon after he came to power in 1979; it included the manufacture of anthrax, plague, botulinum toxin, and aflatoxin, albeit in nothing like the quantities reported for the Soviet Union. Hussein was apparently deterred from using these weapons during the 1991 Gulf War by the threat of overwhelming military reprisals on the part of the United States. In the diplomatic battles preceding the 2003 invasion of Iraq, Hussein's alleged possession of weapons of mass destruction, including bacterial weapons, was a central issue, but none were found either by U.N. inspectors before the invasion or by the U.S.-British forces afterward. Apparently, Hussein had been telling the truth when he said that he had terminated those programs and destroyed existing stocks of biological agents.

In the dark days after the terrorist attacks of September 11, 2001, the American people were put in a state of even greater alarm by a series of small-scale but deadly attacks involving anthrax. Several letters containing anthrax in powder form were mailed to national news organizations and politicians. The intended targets, who included NBC television news anchor Tom Brokaw, Senate Majority Leader Tom Daschle, and Senator Patrick Leahy, escaped harm, but twenty-two other people contracted anthrax infections and five of them died. In August 2002 Attorney General John Ashcroft named Steven Hatfill, a former Army bioweapons researcher, as a "person of interest" in the investigation of the attacks, but Hatfill vigorously protested his innocence and brought lawsuits against Ashcroft, the FBI, and various news media for violation of his rights or for defamation. A lawsuit against the *New York Times* was dismissed in early 2007; other cases are still pending. Five years after the attacks, no charges have been brought against anyone, the perpetrator remains unidentified, and his or her motivation is a matter of speculation.

Another accident involving the airborne release of a virus from a laboratory occurred in England less than a year before the Sverdlovsk incident. The virus was smallpox, and the release was not from a military facility but from the laboratory of Henry Bedson, a smallpox researcher at the University of Birmingham. Janet Parker, a forty-year-old medical photographer who worked on the floor above Bedson's laboratory, apparently inhaled virus particles that had entered her workspace through the building's ventilation system. Parker died—the last person on Earth to be killed by smallpox, which was eliminated in the wild in 1977. A few days after Parker's diagnosis, but before she died, Bedson walked down to the potting shed at the bottom of his garden, slit his own throat, and bled to death.

CHAPTER 9

FORENSIC SCIENCE:
The Wrong Man

On Friday, October 30, 1998, a sixteen-year-old African American youth by the name of Josiah Sutton walked to a neighborhood convenience store at the corner of Fondren Road and West Bellfort Street on the southwest side of Houston, Texas. He was accompanied by his friend Gregory Adams. Although they were behaving innocently enough, the two youths were stopped by police, handcuffed, and placed in the back of a squad car. Sutton didn't see freedom again for four and a half years.

Five days earlier, sometime after midnight on the night of October 25, a forty-one-year-old rape victim arrived at a Houston-area hospital. After medical treatment and the taking of forensic samples, she gave investigators the following account of what had happened: At about eleven p.m. she was parking her SUV in the parking lot of her apartment complex on Fondren Road when she was approached by two African American teenagers, one of whom was armed with a gun. They forced her back into her car and drove it to a deserted location. There, on a bench seat in the rear of the vehicle, they forced her to engage in oral and vaginal intercourse with both of them. The woman said that two other men approached the parked vehicle and witnessed the rape, but they did nothing to help her and in fact chatted casually with the attackers. They even suggested at one point that the rapists should take care not to leave fingerprints. After the rape, the two attackers drove her to the southern outskirts of Houston and left her in a field.

The woman described her two attackers to police. They were black, no more than twenty years old. The one who had held the gun to her head and who had driven her SUV was about five feet seven inches and weighed about 135 pounds. Thus she was describing someone who was significantly smaller than herself—the woman was five feet ten inches and over 200 pounds. This man wore a baseball cap with the bill turned sideways. The other man, the woman said, was about the same height as the first, but even skinnier—about 120 pounds. He had been wearing a skullcap.

After several days spent recovering at another location, the woman returned to her home. On the Friday, as she was driving her car in her neighborhood, she saw Sutton and Adams on their way to the convenience store. She noticed that Sutton was wearing a baseball cap turned sideways, and Adams was wearing a skullcap. Relying in part on these features, she recognized them as her attackers, and she drove to the Fondren police substation to report the sighting. A police officer immediately located the two men and arrested them. Then, while they were sitting handcuffed in the back of a police car, the police brought the woman by. She viewed the men from inside her own car, which was stopped about ten feet from the police car. Again, she positively identified Sutton as the man with the gun, even though Sutton was six feet tall and 200 pounds—five inches and sixty-five pounds more than her previous estimate. She also identified Adams as her other attacker.

The two youths were detained, and the next day they were charged with kidnapping and rape. Both men protested their innocence. They volunteered to give blood samples for DNA testing, and these were taken. It took the Houston Police Department's Crime Lab several months to process the DNA, however. During this waiting period Sutton was ordered to stand trial as an adult, and he was moved from juvenile hall to the county jail.

This arrest wasn't Josiah Sutton's first time in trouble. According to an account in the *Houston Chronicle,* the boy had quite a difficult upbringing. His father left home when Josiah was about six years old and was currently in jail on drug charges. His mother, Carol Batie, had to raise Josiah and his four younger siblings on her own with a low-income job, supplemented later by a small amount of money that Josiah earned by cutting hair. In 1997 Josiah failed his freshman year in high school. In July of 1998 he was arrested for illegal possession of a gun and sentenced to probation, but he violated the terms of his probation by failing to meet with his probation officer. He switched to a new school but was suspended for fighting, and later he dropped out of school altogether. His girlfriend became pregnant. By October, Josiah seemed to be on a downward spiral that could easily have culminated in a serious brush with the law. Nevertheless, during his time in the county jail, Josiah continued to maintain his innocence, and he expressed confidence that he would be exonerated by the results of the DNA analysis.

Meanwhile, in the Crime Lab, the DNA study was being done by Christy Kim, who had twenty years' experience as a forensic analyst. Kim had four samples that might contain the perpetrator's DNA. These were a swab from the victim's vagina, combings from her pubic hair, a stained portion of her jeans, and a stain—identified as semen—from the bench seat of her SUV where the rape had occurred. In addition, she had blood samples from the victim as well as from the two suspects, Sutton and Adams. Thus, Kim's first task was to determine whether the DNA present in Sutton's and Adams's blood samples matched that found in any of the four evidentiary samples. If it did, her second task was to determine, using statistical techniques, how strongly the match pointed to Sutton or Adams as the perpetrator.

Kim used a standard procedure to perform the analysis. First, she extracted DNA from the samples. In the case of the vaginal sample, she performed chemical procedures to isolate any sperm DNA that might be present, thus removing the victim's own DNA. Next she amplified the DNA in the various specimens, meaning that she increased the amount of DNA in the sample by using the polymerase chain reaction (PCR) technique. Because the PCR technique can amplify even a very few molecules of DNA, it was necessary to take care that each sample was not cross-contaminated with DNA from the other samples or from other sources within the Crime Lab.

Finally, Kim did the actual DNA typing. Using commercially available test kits, she examined seven different sites in the genome (called loci), at each of which there are two or more different DNA sequences (or alleles) present in the population. These differences consist of varying numbers of repeats of a short DNA sequence: One allele might consist of three repeats, another of four repeats, and a third of five repeats, for example.

When six or seven loci are tested, the odds that an innocent suspect's alleles will match those of the perpetrator merely by chance are very small—the odds are at least a few hundred thousand to one against this happening, and they may be much slimmer than that. Since there are only about 60,000 black teenagers in Houston, these kinds of odds made it unlikely in principle that an innocent person would be misidentified as the perpetrator of a crime such as the one Sutton was accused of.

There is one slight complication. It's common for an individual to possess two alleles rather than just one allele at a given locus. That's because people possess two versions of each chromosome—one inherited from their mother and one from their father. It may happen that the alleles on these two chromosomes are the same, but quite commonly they are different. If a suspect has two different alleles at a given locus, those two alleles should be present in the evidentiary sample also, if the suspect is the actual perpetrator.

When Christy Kim examined the test results from the rape case, she immediately made a couple of important observations. With regard to the sperm sample from the victim's vagina, the printout showed *more* than two alleles at some loci. Since no single individual possesses more than two alleles at a given locus, the sperm sample had to contain a mixture of DNA from more than one individual.

By itself, this finding was consistent with the victim's account of having been raped by both Sutton and Adams. However, Kim also saw that, at several loci, Greg Adams possessed at least one allele that was not present in the vaginal sperm sample. Thus none of the sperm found in the victim's vagina came from Adams. The sperm sample on the seat of the car also failed to match Adams's DNA.

Of course, Adams might have raped the victim without leaving sperm in her vagina or on the car seat. But if so, only *one* man's DNA should have been present in the vaginal sample, rather than two, because the victim stated that she had not had intercourse with any men other than the two rapists within the previous week. Given these facts, the DNA findings ruled Adams out as one of the rapists.

Evidently the victim had misidentified Adams as one of her attackers. Perhaps she had relied too strongly on that one item of identification, the skullcap, which is after all a popular form of headgear among African American teens. When Christy Kim passed on the results of the DNA analysis to the prosecutor's office, the D.A. quickly realized that the negative DNA findings trumped the positive identification by the victim. He dropped the charges against Adams, who was released. Nevertheless, he had been in custody for five months before this happened.

Adams wasn't the first person to suffer in this fashion: In 1996 the *Houston Chronicle* reported that another sexual assault suspect was held for nearly nine months while the Crime Lab performed a DNA analysis that ultimately proved him to be innocent. Yet tests of this kind can be completed in two to three days.

For Sutton, the outcome of the DNA testing was quite different. With regard to the sperm sample taken from the victim's vagina, Kim found that Sutton's alleles were present at all seven loci that the machine tested. (There were also other alleles present that didn't match Sutton's alleles, but that was to be expected: They presumably belonged to the other, unidentified attacker.) And, Kim reported, Sutton's alleles were also present in the sample of pubic hair combings, as well as in the sperm sample retrieved from the car seat where the rape had occurred.

Kim then went on to calculate how common Sutton's DNA profile was. She did this by consulting reference tables that listed the prevalence of the various alleles at each locus for the demographic group that Sutton belonged to. She calculated that Sutton's pattern of alleles occurred in only one in 694,000 black males. Josiah Sutton, it seemed, had left his genetic fingerprints all over the crime scene.

With positive identifications by both the victim and the Crime Lab, the D.A. was not about to let Sutton walk. He did enter into plea bargaining with Sutton's attorney, but his offer—fifty years' imprisonment—was hardly calculated to encourage a guilty plea, especially while Sutton was still loudly proclaiming his innocence. Carol Batie hired a different attorney, Charles Herbert, for her son. Encouraged by Sutton and Batie, Herbert requested that DNA samples be sent to an independent lab for testing. Kim did send the samples, but somehow they never got tested. By the time the trial got under way, in July 1999, there had been no independent review of Kim's analysis.

The O. J. Simpson trial it wasn't. The entire proceedings lasted just three days. The victim described her ordeal and pointed to Sutton as the man who had abducted her at gunpoint and raped her. Christy Kim described her analysis of the DNA. Herbert and the trial defense attorney made some efforts to challenge the believability of the witnesses. Herbert, for example, pointed out that the Houston Crime Lab was not accredited by the American Society of Crime Lab Directors as mandated by Texas law. But it didn't seem to matter. The prosecutor, Joseph Owmby, described Sutton as "evil and dangerous," and the jury agreed: They took two hours to find him guilty. The judge, a career prosecutor named Joan Huffman, who had just been appointed to the bench, sentenced Sutton to twenty-five years' imprisonment.

In prison, Sutton witnessed murder, rape, and suicide. He was injured in fights and he experienced solitary confinement. But he also converted to Islam, earned his GED, and studied legal issues relevant to his case. He appealed his conviction, but lost. He wrote to the trial judge, asking for retesting of the DNA samples, but Huffman refused.

Sutton also wrote to the Texas Innocence Network—an organization based at the University of Houston. His application for legal assistance was rejected because the Network had a policy not to challenge DNA-based convictions. At that time the Innocence Network, like other similar projects around the nation, viewed DNA testing as the gold standard—the ultimate arbiter of the truth that might get innocent people out of prison but would never put them there in the first place.

The first glimmer of a change in Josiah Sutton's fortune came in the spring of 2001, when the Texas state legislature enacted a revision to the criminal code that allowed convicted criminals to apply for retesting of DNA evidence under certain conditions. Early in 2002 Sutton filed a request for retesting, but his request was just one of hundreds that flooded the system. Nothing much happened except that Sutton was moved back from prison to the county jail.

In the course of 2002 two reporters for KHOU-TV, the local CBS affiliate, began to investigate the Houston Crime Lab. The reporters, Anna Werner and David Raziq, had been alerted to potential problems by defense attorneys. The reporters examined the Crime Lab's records for numerous cases and sent seven of them for review by outside experts. One of these experts was William Thompson, a professor of criminology, law, and society at the University of California, Irvine. Thompson had achieved some celebrity as a member of O. J. Simpson's defense team; he had critiqued the handling of DNA samples in that case.

When I met with Thompson in 2006, he told me that he was shocked when he studied the material he received from Werner and Raziq. "The first seven cases I looked at, there were egregious problems. There were outright misrepresentations of lab results: Analysts would tell the jury that sample A matched sample B, and I would look at the underlying lab work and sample A did not match sample B. And they were failing to run proper control samples. When DNA testing is done, it's important to run a control called a 'reagent blank' [a test with no DNA], in order to make sure that you're not contaminating these samples with foreign DNA. When you're extracting the evidence samples along with the references samples, which Houston was doing, you could accidentally contaminate the bloodstain from the crime scene with the suspect's DNA and thereby get him mixed in. In my own work, I always look closely at the reagent blanks to make sure they're clean. Well, there weren't any reagent blanks. In seven cases I didn't see reagent blanks in any of them."

Another problem had to do with the way the statistics were presented to the juries, especially in cases where the evidence sample contained a mixture of DNA from more than one person. The appropriate statistic to report is the fraction of the population whose alleles are represented in the mixture of alleles found in the evidence sample—that is to say, the fraction of the population who could have donated DNA to the sample. "But they were not providing that statistic," said Thompson. "They were providing the frequency in the population of people who would exactly match the suspect. Most other labs had figured out that in a mixture case you use what are called 'mixture statistics.' So this lab was not following general practices."

On November 11, 2002, KHOU ran the first of a series of investigative news reports about the deficiencies at the Houston Crime Lab, based in part on Thompson's analysis. The reports drew a great deal of attention, because Harris County, where Houston is located, has long had a reputation as the death-penalty capital of the world: It has executed more people since 1982 than any other city or state except Virginia. Many of these cases have involved physical evidence that was analyzed by the Crime Lab.

In response to the television reports and the public outcry that they provoked, the Houston Police Department commissioned an audit of the DNA lab, which revealed numerous deficiencies. In December, the DNA lab was closed down indefinitely.

One of the people who saw the KHOU reports was Josiah Sutton's mother, Carol Batie. She immediately contacted Werner and Raziq and told them about her son's case. They collected the lab reports and court transcripts from the case and sent them off to Thompson for his opinion.

Thompson recounts that he and his wife, Claudia, worked on the boxful of reports together over the breakfast table one Saturday morning. Thompson studied Christy Kim's lab report while his wife read the transcripts. One passage in the report read as follows:

> A mixture of DNA types consistent with J. Sutton, the victim, and at least one other donor was detected on the vaginal swabs, unknown sample #1, debris from the pubic hair combings, and the jeans based on PM, DQA1, D1S80 typing results.

Thompson didn't know what "unknown sample #1" referred to, but by reading the transcript Claudia deduced that it was the semen stain found on the car seat where the rape had occurred. This meshed with what Christy Kim had testified in court—that the semen on the car seat could have come from Sutton. But when Thompson looked at the actual test results, he saw that this couldn't be true: At the locus named "DQA1" Sutton possessed the alleles known as "1.1" and "2," whereas the semen stain contained the alleles "2" and "3." Sutton's allele "1.1" was not present in the stain, and thus he could not have contributed to it.

Kim had actually read the tests correctly, because her lab notes recorded the mismatch between Sutton's DNA and the semen on the car seat. But the report she wrote up for the police ignored the mismatch and wrongly fingered Sutton as a possible source of the semen. Questioned about this later, Kim said that her misstatement was caused by a "transcription error."

The fact that Sutton wasn't the source of the semen on the car seat didn't let him off the hook, of course, since the semen could have come from the other attacker. So Thompson turned his attention to the sperm sample from the victim's vagina. It was true, as Kim had reported, that the DNA from this sample contained Sutton's alleles at every locus tested. At the DQA1 locus, however the sample contained a total of four alleles: 1.1, 2, 3, and 4.1. If the other attacker's alleles were 2 and 3, as indicated by the semen on the seat, Sutton must have contributed the remaining alleles, 1.1 and 4.1, if he was one of the rapists. But that was impossible, because Sutton didn't possess the allele 4.1. "We had this sudden realization over the breakfast table," said Thompson. "'Wait a minute,' we thought. 'This makes no sense!'"

By themselves, Thompson's deductions didn't absolutely prove Sutton's innocence, because they depended on two assumptions. One was that the semen sample on the car seat did indeed come from one of the rapists, rather than being the result of some other sexual encounter that just happened to have taken place at the same location. Also, it was theoretically possible that Sutton was one of the rapists but a third man—neither Sutton nor the person who deposited semen on the car seat—was the source of the unexplained 4.1 allele in the vaginal sperm sample. Still, that seemed unlikely, since the victim said she hadn't had intercourse with anyone other than the two attackers in the relevant time frame. Thus the most likely conclusion was that Sutton was in fact innocent.

Thompson also took issue with Christy Kim's presentation of the statistics. Just as had happened in the previous cases that Thompson reviewed, Kim had simply reported the odds that a randomly chosen black person would match Sutton's alleles at the loci she tested. She gave that figure as 1 in 694,000. She also cited this figure in court. The jury was therefore left with the impression that there were astronomical odds against the possibility that Sutton's DNA matched the vaginal sperm sample purely by chance. Yet, as Thompson demonstrated, the fact that the sample was a mixture of at least two individuals' DNA shortened the odds tremendously. In fact, at three of the seven loci, any person on earth would have matched the vaginal sperm sample, because at each of those loci only two possible alleles exist, and the sperm sample contained both of them. When Thompson used the appropriate procedure to calculate the odds that a randomly chosen black person would match the mixed sperm sample at all seven loci, he came up with a figure of not 1 in 694,000 but 1 in 15. And considering that the police had two shots at finding a match (once with Josiah Sutton and once with Gregory Adams) the odds that one of the men would match the sample were twice as high—1 in

7.5. Thus, even if one ignored the fact that Sutton was positively excluded as a suspect (with the caveats mentioned above), the match of his DNA to the vaginal sample could easily have been a mundane coincidence of the kind that a busy crime lab could expect to encounter every day.

When Werner and Raziq received Thompson's report, they aired Sutton's case in a news special on KHOU-TV. A few days later, the Houston Police Department ordered a new test of the DNA evidence, this time from an independent laboratory named Identigene. The laboratory used a more recently developed test named Profiler Plus, which uses a special-purpose machine, combined with analytical software, to test nine different loci simultaneously. These loci are different from the ones tested by Christy Kim, and they are more informative because there are more possible alleles (7 to 14) per locus. The results proved that two men contributed to the vaginal sperm sample and that neither of them had the DNA profile of Josiah Sutton. Sutton was innocent.

A few days later, on March 12, 2003, Sutton was released on bond. Now twenty-one years old, his life as a free man took an erratic course: He tried various jobs and even enrolled in college, but nothing seemed to stick. Meanwhile, the director of the Texas Innocence Network, David Dow, finally agreed to take on Sutton's case, with the aim of getting the parole board to recommend to Governor Rick Perry that he should issue a pardon.

Texas has two kinds of pardon. A "full pardon" leaves the conviction standing: it simply terminates the sentence and restores the person's civil rights, without saying anything about whether he actually committed the crime. A "pardon for innocence" is much rarer: It annuls the original conviction and opens the door for the person to claim compensation for time spent in prison—at a rate of up to $25,000 per year. To obtain a pardon for innocence the district attorney, the police chief, and the judge must all send letters to the parole board recommending the pardon, and the letters much be accompanied by evidence that the person did not commit the crime for which he was convicted.

In Sutton's case the district attorney, Chuck Rosenthal, seemed incapable of letting go of his belief that Sutton was one of the rapists, and he therefore petitioned for a full pardon rather than a pardon for innocence. When asked by journalists for his reasons, he mentioned the fact that the victim had positively identified Sutton as one of her attackers (though she had also positively identified Gregory Adams, who was unquestionably innocent). As for why Sutton's DNA was not found in the evidence samples, Rosenthal suggested that he had used a condom or failed to ejaculate. This theory required the victim to have been raped by three men rather than the two she testified to. This kind of reasoning—adding extra, unknown perpetrators—has become a staple among prosecutors trying to rescue a case in the face of exculpatory DNA evidence. Peter Neufeld, cofounder of the New York Innocence Project, has famously dubbed it the "unindicted co-ejaculator" hypothesis.

Dow's efforts to obtain the pardon for innocence dragged on for months. Rosenthal initiated yet another independent round of testing of the DNA evidence in the Sutton case, and the results confirmed Identigene's earlier report. Finally, in May 2004, Governor Rick Perry issued Sutton a pardon for innocence. Sutton petitioned for compensation and was declared as being entitled to $118,000. He received his first payment of $60,000 in 2004. He went through the money in six months by buying three cars, partying, and giving money to his family, according to the *Houston Chronicle*. He told the newspaper that he planned to be more careful with the next installment. "It was a learning experience, and I have grown up," he said.

"After he was exonerated, I went down to testify before a grand-jury hearing about the lab," said Thompson, "and I took him and his mother out to dinner. I was kind of relieved to meet him, because I was thinking he would be some thug, but he struck me as sort of a gentle soul. He's into meditation; he's very serious about religion. He seemed confused and rootless and he didn't quite know what to do with himself."

While Sutton was seeking his pardon, the city of Houston turned its attention to the Crime Lab. In early 2003 the former police chief recommended that the lab's director be fired, and he retired before that could happen. Nine employees of the lab were disciplined. Kim and another analyst were given fourteen-day suspensions, but they appealed to the Civil Service Commission and the suspensions were overturned. Then, in December, the mayor fired Kim, but again the Commission reinstated her, on the grounds that she had merely followed the laboratory's standard practices. Eventually Kim resigned, and several other employees joined the director in retirement.

In early 2005 the city of Houston commissioned a new and completely independent review of its Crime Lab by Michael Bromwich of the Fried Frank law firm. As U.S. Inspector General during the Clinton administration, Bromwich had conducted an investigation of the FBI's Crime Lab that led to the disciplining of several agents.

Bromwich's review of the Houston lab, published in several installments in 2005 and 2006, was a damning indictment of the lab's practices, not just in DNA work but in serology (blood-group analysis) and other departments. Out of sixty-seven DNA cases reviewed, Bromwich found that 40 percent had deficiencies serious enough to raise doubt as to the reliability of the work, the validity of the results, or the correctness of the analysts' conclusions. Three of these were capital-murder cases that ended with the defendant being convicted and sentenced to death, though none had proceeded to actual execution.

In numerous instances, the analysts seemed to have shaded their findings to support the prosecution, often by failing to report findings that contradicted the prosecution's case. In one serology case, for example, a bloodstain found at the scene of a double murder failed to match either the victims or the accused—a finding that was clearly exculpatory—but the DNA lab chief, James Bolding, concealed the mismatch by marking the stain's blood type as "inconclusive," according to Bromwich's report. Years later, a second person was accused in the case, and his blood did match the stain. According to Bromwich, Bolding did not retest the stain but simply altered the original report to show a positive match. Bromwich also wrote that Bolding was even untruthful about his own qualifications: In a sexual assault case he testified that he had a Ph.D. in biochemistry from

Texas Southern University, but he later admitted to Bromwich that he did not have a Ph.D. degree at all.

There were also DNA cases in which the analyst reported having run controls (such as reagent blanks) but the lab data showed that no controls had been run. Such cases of apparent "dry-labbing" are said to be pervasive in crime labs around the nation.

The analysts consistently misreported the statistics of their cases. Just as happened in the Sutton case, the analysts would cite the frequency of the accused's DNA profile in the population (a very low but irrelevant number), but not the likelihood that the accused's profile would be present in a mixed evidence sample purely by chance (a much higher number). "It is clear that the DNA analysts in the Crime Lab, including Mr. Bolding, did not fully understand the scientific basis of calculating frequency estimates," wrote Bromwich.

Perhaps the overriding problem at the Houston Crime Lab was the lack of effective oversight. "We have found no semblance of an effective technical review program or quality assurance regime to detect and correct these problems," wrote Bromwich. "As a result, they continued unabated."

Bolding eventually ceased cooperating with the investigation, and Kim had refused to answer Bromwich's questions from the beginning. Although this lack of cooperation may have been motivated by a fear of self-incrimination, no Crime Lab employees have in fact been indicted in connection with the investigations.

Although many convictions have been brought into question by Bromwich's review, so far only two persons have been released from prison. One was Sutton, and the other was a man who served more than seventeen years for rape and who, like Sutton, was exonerated by new tests.

In the summer of 2006, after being shuttered for three years, the Houston Crime Lab reopened for business with official certification, a new director, new or retrained analysts, a quality-control program, and largely new facilities. A year later, Bromwich's final report spoke approvingly of these changes, but also listed some persisting shortcomings in the techniques used to analyze and interpret DNA data.

In fact, there are considerable differences of opinion about the nature of the underlying problems and how to solve them. According to Bill Thompson, the core problem is that police crime labs are instruments of the state, so their analysts will consciously or unconsciously favor the prosecution. The way to solve this problem, according to Thompson, is to close the police labs and turn all work over to independent agencies. In theory at least, these labs would be more likely to conduct scientifically rigorous tests and to present the results in an impartial manner.

In addition, Thompson believes that all labs should be subject to periodic random testing in which known samples are provided for "blind" analysis—that is to say, the laboratory would not know when they were being tested. Currently, if such quality-control tests are done at all, analysts are given advance notice. As a result, we may assume that they are put on their best behavior. Thompson believes that blind testing could be used to derive some measure of a lab's false-positive rate—the likelihood that an incriminating test result would be in error. Even if the false-positive rate were very low—say, one in a thousand—that would still make nonsense of the one-in-a-billion-type statistics that are commonly presented to juries in an effort to secure conviction.

A committee of the National Research Council has rejected these ideas. It would be impracticable and too expensive to conduct enough blind testing to obtain reliable estimates of false-positive rates, the committee believed. And if defense lawyers believed their case was affected by pro-prosecution bias among police-lab analysts, they should be encouraged to commission independent testing to counter it.

That's not realistic, counters Thompson. The defense attorneys may not be familiar enough with DNA technology to perceive the need for a retest. There are often insufficient funds for independent testing. Or the police lab may have used up the entire evidence sample in their own testing—or they'll say that they have.

Most importantly, criminal defense lawyers are often reluctant to commission independent tests for fear of harming their clients.

After all, defense lawyers generally assume that their clients are guilty—because they usually are. So they expect that a test conducted by an independent lab will confirm the findings of the police lab. For sure, they don't have to mention in court that they had their own testing done or what the result was, but the prosecution often finds some devious way to bring up the subject, according to Thompson. "The prosecutor says to the expert, 'Oh, I notice there are some missing pieces of this fabric sample, where did those go to?' The defense says, 'Objection!' and the prosecutor says, 'Your Honor, it's important for continuity, to explain, etcetera.' The judge says, 'Objection overruled,' and the expert's answer is 'Oh, I sent them to the defense lab for testing.'" Once the jury knows that the defense had its own testing done but didn't present the results, they immediately assume that the results were incriminating, so any attack on the police lab's tests loses credibility. Thus there are several real-life factors that reduce the defense's ability or willingness to conduct independent tests.

Thompson's self-appointed mission is to educate defense lawyers about DNA testing so that they are in a better position to review tests conducted by police labs and spot the problems that call out for independent testing and for challenging the state's experts. The lawyers and law students who staff the Innocence Projects are clearly hearing his message, because the existence of DNA evidence is no longer a bar to taking on a case the way it was when Sutton sought the Houston Innocence Network's aid.

Have errors in forensic science ever led to the execution of an innocent man? Given the sheer number of executions—more than a thousand since the death penalty was reinstated in the United States in 1976—it seems likely that they have, but identifying a specific instance has proved difficult. One case that has drawn a lot of attention is that of Cameron Todd Willingham, executed in Texas in 2004 for setting a house fire that killed his three children. Willingham asserted his innocence before, during, and after his trial, and he did so again in his final statement before execution. At his trial, investigator Manuel Vasquez reported finding scientific indicators of arson, such as the presence of crazed glass, but subsequently published forensic guidelines have rejected them as mere superstition. Hot glass is easily crazed by contact with water from fire hoses, for example. "Each and every one of the 'indicators' listed by Mr. Vasquez means absolutely nothing," reported a commission of nationally respected arson investigators in 2006. Willingham certainly hasn't been proven innocent, but the evidence for his guilt has largely evaporated.

Josiah Sutton's story isn't over. When the independent lab did the DNA testing that ruled him out as a suspect, it was able to reconstruct a complete DNA profile for one of the actual rapists, as well as a partial profile for the other. The complete profile was used to search a database maintained by the Texas Department of Public Safety, but no match was found. Still, new DNA profiles are being entered into the database all the time, because in Texas all convicted felons have to give a DNA sample. In 2005 a young black man named Donnie Lamon Young, who was serving time on a drug conviction, gave blood for testing. In May of 2006 the DPS found that Young's DNA was a match to the complete profile from the Sutton case.

The Houston Police Department was notified, and a new sample was taken from Young, who was by then out of prison. Again there was an exact match. In June, Young was arrested and charged with aggravated sexual assault in the case for which Sutton was wrongly convicted. He was held in the county jail after he failed to post a $150,000 bail bond.

The victim was unable to pick Young out of an identification lineup, but in January 2007, Young pleaded guilty, and he was sentenced to ten years imprisonment. He also named his accomplice, a man who had died in prison.

When asked by the *Houston Chronicle* for his reaction to Young's arrest, Josiah Sutton expressed himself laconically: "Let's just say, if he's the one who did it, that I don't think we would be two good people to put in a room together." His mother was more philosophical:

"My son had been pardoned," she said, "but it still weighed on my heart that no one had been arrested and that some people would not believe in Josiah's innocence until someone was. Now, justice can be done for the victim, and we can really close the book and say he did not do it."

District Attorney Rosenthal said, "I still don't know enough to know whether [the victim] was mistaken or not. I intend to look into it, but if Sutton is innocent, I will be the first to say he is." To which Bill Thompson commented, "Chuck, you're a little late!"

CHAPTER 10

SPACE SCIENCE:
Off Target

On September 23, 1999, after a 419-million-mile journey, the Mars Climate Orbiter spacecraft made its final approach to the Red Planet. At one minute after two in the morning, a tired but excited group of engineers and scientists at NASA's Jet Propulsion Laboratory (JPL) near Pasadena, California, broke into smiles and applause: A signal had arrived, indicating that the spacecraft's main engine had begun firing. This event would reduce the spacecraft's speed enough for it to be captured by the planet's gravitational field and go into orbit. Shortly thereafter, as expected, signals from the spacecraft ceased as it passed into the radio shadow behind the planet. The JPL group waited impatiently for the spacecraft to reemerge on the other side of the planet, an event that was predicted to occur twenty-one minutes later. But the silence stretched to twenty-two, twenty-three, and twenty-four minutes, and then to hours and days. In fact, no signal was ever received from the spacecraft again. The Mars Climate Orbiter was lost, and the mission was a total failure.

Losing a Mars mission was not exactly a new experience for NASA: Three out of the eleven previous U.S. missions had ended in failure. Just six years before the Mars Climate Orbiter mishap, the $800-million Mars Observer spacecraft was lost in rather similar circumstances: Radio signals from the spacecraft mysteriously ceased during its final approach to Mars.

Still, the U.S. Mars program overall had a good track record, especially in comparison with the Soviet program. Of the eighteen

Russian spacecraft sent to Mars before 1999, fifteen had been total losses, often failing to reach space at all. The remaining three were only partial successes. And of the successful earlier American missions, some had been extraordinarily complex. These included the *Viking* mission of 1975—a fleet of two orbiters and two landers, all of which functioned as planned and for far longer than their nominal design lifetimes. By the time of the Mars Climate Orbiter launch, there was a real confidence that NASA and its industrial partners—Lockheed Martin Astronautics, in this case—knew how to get the job done. It may have been this very confidence that sank the mission.

The root cause of the loss was a scientific blunder as old as science itself: the confusion of units. Such errors can be prevented. Yet in a larger sense the loss represented a failure in systems engineering—that is, a failure to successfully integrate thousands of individual technical contributions into a single cohesive whole: a product that would fulfill the objectives of the customer, the U.S. government. In that sense, the Mars Climate Orbiter mishap represented what is probably the most common mode of failure in large and complex scientific enterprises, and one that is extremely difficult to eradicate.

In the early 1990s, NASA administrator Dan Goldin spearheaded a new approach to the design and implementation of space missions, an approach that was encapsulated in the slogan "Faster, Better, Cheaper," or FBC. The FBC philosophy was in part a response to fiscal belt-tightening imposed by the U.S. government. It also represented a reaction to some of the earlier missions: huge, long-delayed projects that incorporated every imaginable bell and whistle and that went tens or hundreds of millions of dollars over budget. FBC was a leaner approach that aimed to achieve more with less, employing economical strategies such as the reuse of design elements that had proven successful in earlier missions. Although this "heritage" approach promised great savings in time and money, it also injected risk. How could one be sure that a large piece of software, for example, would function successfully in the different environment of a new spacecraft? And the FBC approach also demanded economies in manpower. This was something that might be acceptable so long as things went according to plan, but it might cause problems when unexpected difficulties needed to be surmounted, as happened with the Mars Climate Orbiter.

Goldin's strategy had some early successes. The 1996 Mars Pathfinder mission, for example, safely delivered the rover *Sojourner* to the Martian surface using a novel airbag landing system. The rover was able to navigate semiautonomously around the landing site and it did some simple geological investigations of nearby rocks. It also caught the imagination of the public, including children, back on Earth: Mattel's rover action model was the hottest toy of the summer of '97.

Soon after Pathfinder's landing another spacecraft, the Mars Global Surveyor, reached the planet and went into a polar orbit. Over the following ten years it took nearly a quarter-million photographs of the Martian surface, and it also operated as a communications satellite for other missions until it ceased functioning in 2006.

In spite of these successes, there were also hints of problems with the FBC approach. In 1997, for example, an Earth-orbiting satellite called Lewis (as in Lewis and Clark) was launched, but once in orbit it went into a spin that prevented its solar panels from facing the sun; this caused the batteries to lose charge. The problem occurred at night while the controllers were off duty; economic considerations had prevented the appointment of sufficient controllers for round-the-clock staffing. By the time the controllers returned to work the next morning, the spacecraft was completely out of electrical power and thus could not be resuscitated: It burned up in the atmosphere a few weeks later.

The Mars Climate Orbiter (MCO) was one element in a two-spacecraft mission named the Mars Surveyor '98 Program. The other element was the Mars Polar Lander. The role of the Climate Orbiter was to study the Martian atmosphere with a variety of instruments and also to serve as a communication link for the Lander and for other, future missions. The

Lander was to set down near the planet's south pole—using retrorockets rather than an airbag—and dig into the soil with the particular aim of finding water.

Lockheed Martin won the $121-million contract to build both spacecraft (not including the scientific instruments). The company was expected to complete the job with minimal oversight from NASA, consistent with the Faster, Better, Cheaper philosophy.

The error that doomed the MCO spacecraft occurred during the development of its navigational software. To understand the error, it's necessary to appreciate that once a Mars-bound spacecraft has escaped Earth's gravitational field, it is essentially coasting in an orbit around the sun, albeit a highly elliptical orbit that is carefully planned to intersect the orbit of Mars at a time when Mars itself has reached that location. Thus, if the spacecraft's initial direction and speed are well enough known, its future trajectory can be readily calculated from gravitational equations provided by Isaac Newton.

Two nongravitational factors can affect the trajectory, however. One consists of deliberate changes in trajectory induced by firing of the spacecraft's small rocket engines, or "thrusters." Usually, four or five of these trajectory correction maneuvers (or TCMs) are performed in the course of the flight. Each time a TCM is performed, its effect on the trajectory has to be determined.

The other main source of nongravitational effects comes from the pressure of solar radiation on the spacecraft, and from the craft's efforts to compensate for that pressure. Radiation pressure is exerted mainly on the craft's solar panels, on account of their large area. Unlike the Mars Global Surveyor, which had solar panels on either side of the spacecraft, the Mars Climate Orbiter had all of its panels on one side of the craft. Because of this asymmetrical design, the main effect of solar radiation was a tendency to spin the spacecraft around on its axis. Such a spin was undesirable because it reduced the amount of sunlight received by the panels. To counteract this effect, a set of small reaction wheels resembling the metal discs in gyroscope toys were automatically spun up by electric motors. These spinning reaction wheels generated an equal and opposite rotational force, so no actual rotation of the spacecraft occurred and it maintained a constant orientation to the sun.

Of course, the reaction wheels could only be spun up to a certain limiting speed, which was about 3,000 rpm. Because of the spacecraft's asymmetrical design, it took less than twenty-four hours of flight for the reaction wheels to reach this limit. Then their accumulated angular momentum had to be "dumped." This was done by slowing the wheels while at the same time compensating for the effect of the slowing by firing some of the thrusters. In this way the reaction wheels could be brought back to a stop while keeping the spacecraft's attitude constant. Then the cycle began again. These dumps were formally named Angular Momentum Desaturation events, or AMDs, and they occurred about ten times a week during the trip to Mars.

If these thruster firings had only affected the spacecraft's rotation, they would not have disturbed its trajectory through space and would therefore have been irrelevant from a navigational standpoint. Unfortunately, design considerations led to the positioning of the thrusters in such a way that the AMD would also kick the spacecraft sideways by a small amount, and the frequent repetition of these events over the course of the flight would cause the spacecraft to deviate by a significant degree from its planned trajectory—enough to prevent the craft from entering the Mars orbit correctly.

The Lockheed engineers knew about this issue. They also knew that it would be difficult to measure the effect of the AMDs on the spacecraft's trajectory at the time the AMDs occurred. This was because only limited information would be available about the spacecraft's position, speed, and direction of travel during the flight. In general, it is possible to accurately measure a spacecraft's distance from Earth (based on the time for a radio signal to travel from Earth to the spacecraft and back) and its speed along the line of sight from Earth (based on Doppler changes in a fixed-frequency signal emitted by the spacecraft). It is not possible to directly measure its position or speed in the two dimensions that are perpendicular to the line of sight. These other variables can eventually be determined by making repeated measurements of the spacecraft's range and line-of-sight velocity as it follows its elliptical trajectory, but these determinations are slow and subject to various forms of error. Unfortunately, the effects of the AMD events were exerted largely in the difficult-to-observe dimensions perpendicular to the line of sight from Earth.

To solve this problem, the engineers followed a strategy that had been employed in previous missions such as Mars Global Surveyor: They would use a form of "dead reckoning." This depended on knowing exactly how much thrust would be exerted on the spacecraft—and in which direction—when each thruster was fired for a known period of time. With this information, it would be possible to calculate how much the speed and direction of the spacecraft would change during each AMD event, even without measuring those changes directly.

To make this possible, the subcontractor who manufactured the thrusters sent paperwork to Lockheed Martin that documented how much thrust was generated when each thruster was fired. This manufacturer was accustomed to working in English units (pounds, feet, and so on). NASA generally requires the use of metric units throughout its operations and those of its contractors, but it makes exceptions in cases where ordering a change of units may be unduly burdensome, or where it increases the risk that the contractor will make some kind of mistake. NASA made an exception of this kind for this subcontractor, so the paperwork received by Lockheed Martin listed the thrusters' performance in units of pounds of force.

The root cause of the Mars Climate Orbiter mishap was the failure to convert these English units to metric units of force—newtons—in the preparation of a navigational software file called "Small Forces." The purpose of this file was to determine how strongly each AMD event would push the spacecraft out of its intended path. Because the remaining navigational software assumed that the output of the Small Forces file was in newtons, it underestimated the deflection of the spacecraft's trajectory caused by each AMD event by a factor equal to the ratio between pounds and newtons—which is to say, by a factor of 4.45.

To understand how this error occurred, I spoke with John Casani, the onetime chief engineer at JPL who led the lab's internal investigation into the mishap. I also spoke with Steve Jolly of Lockheed Martin, who was the lead systems engineer for the Mars Climate Orbiter. Casani's and Jolly's accounts agreed in one respect: They both told me that the failure to convert the units was primarily the fault of a young engineer who had only completed college a couple of months earlier and who was a new hire at Lockheed Martin. (This person has never been identified by name.)

Casani and Jolly gave me somewhat differing accounts of how the young engineer actually came to make the mistake. According to Casani, the engineer was given the documentation from the thruster manufacturer that contained performance data in pounds, as well as a set of instructions from JPL that specified that the output of the Small Forces file should be in newtons. The engineer simply failed to read the JPL document with sufficient care and thus overlooked the requirement for the conversion. According to this account, then, the engineer thought he was doing the right thing by providing the output in English units.

According to Jolly (who presumably would have been more knowledgeable about the matter), the engineer *did* know that pounds needed to be converted to newtons. The reason he failed to make the conversion, Jolly told me, had to do with the "heritage" issue. The Mars Global Surveyor had used similar navigational software, and the plan was to save money by having the Climate Orbiter "inherit" or reuse it. The new spacecraft used different thrusters, however, so the portion of the software relating to thruster performance was excised, and the engineer's task was to replace that portion with code incorporating performance data for the new thrusters. Unfortunately, he assumed that the code that made the conversion of units was left in the unexcised Global Surveyor software, whereas in fact it was in the excised portion. The conversion was represented simply by the number 4.45 in an equation, without any comment as to its purpose, so it was easy to miss. Thus in writing the new code the engineer left the units in pounds, thinking that the required conversion would be made by the preexisting software.

Although the engineer's mistake was the root cause of the mishap, such mistakes are inevitable as long as science is done by humans.

The more serious error was the failure of anyone to spot the mistake. Part of the problem was that a factor of 4.45 is not a terribly large error in engineering terms: The faulty code produced output that looked quite reasonable. In fact, if the Mars Global Surveyor (with its symmetrical solar panels) had incorporated the same error, that mission would probably not have been affected. It was only the asymmetrical design of the Climate Orbiter, with the resulting need for numerous AMD events, that allowed the small individual errors to accumulate to a mission-endangering level.

Following standard procedures, the faulty software was reviewed, but the error wasn't spotted. Then it went through formal testing: Using fictional AMD events, the output of the software was compared with the output of manual calculations. Unfortunately the manual calculations somehow incorporated the same error as was present in the faulty software, so the two outputs were in agreement and the software was judged to be good.

The Small Forces software was not actually loaded into the spacecraft's computer; rather, it was placed in computers that remained on the ground. The idea was that every time an AMD event occurred, the spacecraft would radio back data about the length of firing of the thrusters, the attitude of the spacecraft, and so on, and the navigational team would then feed the data into the ground computer to extract measures of the magnitude, duration, and direction of thrust. These measures would then be used to adjust the model of the spacecraft's trajectory.

By a bitter irony, the spacecraft's own computer did in fact possess software to make this calculation independently, and these files correctly specified the resulting thrust in newtons. "You can imagine how many times I wake up at night thinking about that," said Jolly. In fact, the spacecraft was even programmed to radio the output of these calculations to the ground, but the navigators did not know this, so no one looked at the incoming data packets or compared them to the output of the erroneous calculations being performed on the ground. If they had done so, the error would have been quickly detected. Even when I spoke with him in 2006, after NASA's official inquiry had established and published the fact, Jolly said that he didn't know that the spacecraft had been transmitting the correct data to Earth.

On December 11, 1998, the Mars Climate Orbiter was launched from Cape Canaveral Air Station in Florida, atop a Delta II rocket. The Delta II is a relatively inexpensive, medium-powered launch vehicle. So as not to exceed the Delta II's lifting capacity, the mission planners had to economize on the weight of fuel carried by the spacecraft—fuel that was required for slowing the spacecraft when it reached Mars. The planners took two steps to save on fuel. First, they sent the spacecraft by a long route that took it more than halfway around the sun: This ensured that it was traveling relatively slowly as it approached Mars, but it lengthened the trip to nine months rather than the six months needed for a more direct route. Second, they planned to accomplish some of the slowing by aerobraking—repeatedly dipping the spacecraft into Mars's outer atmosphere on successive orbits after the first encounter—rather than relying entirely on the spacecraft's engines. Even with these measures, the orbital insertion burn would have to slow the spacecraft by nearly 5,000 kilometers per hour, a task that would consume nearly 300 kilograms of fuel—almost half the total weight of the spacecraft.

The launch went flawlessly. The Delta II's first two stages lifted the spacecraft into low Earth orbit, then the third stage booster rocket fired for 88 seconds, kicking the spacecraft out of Earth's gravitational clutches. After the booster separated, the spacecraft deployed its solar panels and began its long, unpowered cruise toward Mars.

Teams at JPL and Lockheed Martin, led by JPL flight operations manager Sam Thurman, monitored and controlled the spacecraft during its journey. Part of the team was a group of four JPL navigators, led by Pat Esposito, whose task was to determine the spacecraft's trajectory and calculate the required corrections during the flight. The team was also responsible for two other spacecraft, however—Mars Global Surveyor (which was orbiting Mars) and Mars Polar Lander (which was launched on January 3). Only one team member, Eric Graat, could give his undivided attention to the Mars Climate Orbiter. This was a low level of staffing compared with previous and subsequent missions: The successful Mars *Odyssey* mission of 2001, for example, boasted a fifteen-member navigation team. Although neither Esposito nor Graat agreed to speak with me, Sam Thurman told me that they were very much overworked.

The first trajectory correction maneuver (TCM-1) took place ten days after launch. It corrected a deliberate mis-aim in the launch trajectory—a mis-aim whose purpose was to ensure that the third-stage booster did not strike Mars and contaminate the planet with terrestrial germs. The maneuver involved an elaborate sequence of operations. First, the solar array was folded and locked against the spacecraft body to protect it from damage, then the entire spacecraft was rotated so that the firing of its aft-pointing thrusters would deflect the craft's trajectory in the right direction, and then the thrusters were fired for a few minutes to achieve the correct trajectory. Finally, the spacecraft was rotated back into its flight orientation and the solar panels were deployed once more. A second, much smaller trajectory maneuver (TCM-2) was performed on January 26, 1999, and it, too, went according to plan.

About every seventeen hours during the flight, the spacecraft automatically performed angular momentum desaturation (AMD) procedures, firing its thrusters for a few seconds to allow the reaction wheels to be decelerated. The navigators had not been expecting the AMD events to occur so frequently, because when they came onto the job they were not familiar with the Orbiter, and they didn't realize that its asymmetrical design would cause an increased tendency to spin under the influence of solar radiation.

During the first four months of the flight, the navigators did not use the Small Forces software to calculate the effects of the AMDs on the spacecraft's trajectory. This was because the software not only contained the units error (which no one was aware of), but also some other bugs that *had* come to light. Because the tiny effects of the AMD events would only really be important for the final approach to Mars and orbital insertion, the navigators simply did without the output of the Small Forces software, planning to incorporate the data at a later time.

Finally, in mid-April, the ground software was delivered and put into operation. Now the effects of the AMD events (including those that had already taken place) were incorporated into the navigational calculations. But for each AMD event the software told the navigators that the spacecraft had been deflected by an amount that was nearly five times larger than what had actually occurred, thanks to the poison pill that was the units error. Still, each individual navigational solution looked good, because the error was in an unobservable dimension perpendicular to the line of sight.

Only over time, as more and more solutions were calculated along the spacecraft's curving path, did Graat become aware that the individual solutions didn't quite mesh together to form a coherent trajectory. And calculations of the spacecraft's current position that were derived from different data sets (e.g., those based on range or Doppler measurements, or those that were based on different parts of the spacecraft's trajectory) gave a fuzzy cluster of solutions instead of a single, unanimous answer.

Graat discussed this navigational problem with the leader of the navigational team, Pat Esposito. According to John Casani, the problem should have been entered as a formal written record known as an Incident, Surprise, Anomaly form—or ISA—which guarantees that a problem is followed up to a satisfactory resolution, but that's not what happened. "The navigator here at JPL sent an e-mail message to someone at Lockheed Martin, saying, 'Take a look at this, there's something funny going on that we don't understand.' That never got entered into the formal record, which is our normal practice. So someone received this at the Lockheed end and said, 'I'm going to work on this,' and then he got some other task that came along that either he thought was a higher priority, or his boss thought was a higher priority, and he got deflected, and this problem that was communicated to him by e-mail just fell off the table, so to speak. If the form had been filled out, that could not have happened."

Sam Thurman clued me in to the "other task" that got higher priority. A serious incident occurred during the third trajectory correction maneuver, which was performed on July 23. Although the procedure for TCM-3 was the same as for TCM-1 and TCM-2, the process of retracting and locking the solar panels in preparation for the burn went awry. This procedure involved rotating the panels around a ball joint, using a gimbal drive. "There are devices on this gimbal that read out its angular position," he said, "and there were some calibration errors on those things, so the solar array scraped up against the side of the spacecraft and nearly got jammed in the stowed position. That put the spacecraft into 'safe mode': When it tried to unstow after the maneuver, the array wouldn't move, so the software stopped and called up the ground and said, 'Hey, I'm trying to move this and it's not moving; there's something wrong.' So we spent most of the month of the approach phase scrambling to try to resolve this problem with the gimbal drive—because we knew that when we got to orbit insertion we had to have the solar array in the stowed position when we fired the main engine. The support that held the array wasn't strong enough to take that force without the array being stowed. So we knew following TCM-3 that we had a problem we must fix or orbit insertion would fail. And that was very scary—that took a hell of a lot of effort from the team, the spacecraft team [at Lockheed Martin] in particular. So I think the navigation team's problem was they were calling up the spacecraft team, saying, 'Gee, we're seeing this funny stuff, can you help us work the Small Forces modeling and try to understand it?' And they said, 'Oh my God, we've got this huge other problem that could end the mission if we don't fix it in the next two weeks.'"

With TCM-3 out of the way, most of the team members turned their attention to solving the problem of the balky gimbal drive while the navigators prepared for TCM-4—the last scheduled course correction. This was planned for September 15, just eight days before arrival at Mars. TCM-4 was the really crucial maneuver: It had to leave the spacecraft aiming for a point of closest approach to Mars (the point known as first periapse) that was about 150–200 kilometers above the planet's surface. Much higher than that, and the ensuing aerobraking procedure would take too long for the Orbiter to be in place by the time the Lander arrived on December 3. Much lower, and the spacecraft might be damaged by frictional heating in the planet's outer atmosphere; the lowest survivable altitude was thought to be about 80 kilometers. Previous missions had achieved their preset altitudes with extraordinary precision, sometimes missing by as little as four kilometers—not bad marksmanship after a 400-million-kilometer voyage. But the fuzzy navigational solutions made such a precise result unlikely with Mars Climate Orbiter. In fact, the solutions obtained by use of Doppler measurements and those obtained by range measurements were predicting fly-by altitudes that differed by tens of kilometers. No one knew which set of calculations was more accurate, so it wasn't clear exactly how large the TCM-4 correction should be. The team decided to aim for an altitude of 226 kilometers, a height that left considerable leeway in case the spacecraft was coming in lower than the navigators realized.

By September 15 the spacecraft team had the gimbal-drive problem fixed, and TCM-4 went smoothly: The solar panels stowed themselves and redeployed without incident, and the spacecraft did not go into safe mode. Now, with arrival at Mars just a week away, most of the spacecraft team turned their attention to making preparations for the aerobraking phase that would follow insertion into Mars orbit. These preparations were far behind schedule on account of the gimbal problem.

Meanwhile the navigators worked almost nonstop to refine their trajectory calculations. As the days passed, the predicted altitude of the first periapse gradually decreased to about 150 kilometers—a safe altitude, but only if their prediction was correct to within a few tens of kilometers. During the final two or three days before arrival, the navigators got a serious case of cold feet, and they brought up the possibility of making yet another, fifth course correction to raise the spacecraft's fly-by to a safer altitude.

This was the last thing that Sam Thurman, the flight operations manager, wanted to hear. The possibility of conducting a TCM-5 was written into the mission's contingency plans. However, the optional TCM-5 was mainly planned for a different circumstance, namely for a situation in which the *second* periapse (on the first aerobraking orbit) would be at the wrong altitude. Reprogramming a TCM-5 to alter the *first* periapse altitude would be a major task for a relatively small spacecraft team, especially when it was so far behind on its preparations for the aerobraking phase.

"I remember that the nav team chief was nervous," Thurman told me. "He said, 'I think we ought to bump up [the altitude],' and other people [were] saying, 'If we do that, it's going to screw up our aerobraking preparations.'" Thurman himself sided with the latter group. "[Doing a TCM-5] would have put at risk our ability to get into the correct science mapping orbit a few weeks later," he said. "It would have jeopardized our ability to transition people over to preparations for the Lander's arrival. We needed the same people at Lockheed Martin to do that as well as to get aerobraking started."

Given the lack of consensus, it was left to Thurman (perhaps in conjunction with other leaders of the mission) to make the decision, and they decided against a course correction. "I think that was an error of judgment," John Casani told me, "but that's easy to say."

Up until about noon on the day before orbit insertion it looked as if Thurman had made the right judgment call, because the navigational solutions began to cluster more closely together, suggesting that the 150-kilometer periapse prediction was correct. Unfortunately, the solutions were clustering tightly around an incorrect value. At about one a.m. the following morning, which was an hour before the spacecraft's arrival at Mars, a new set of solutions became available, refined by observation of the ever-increasing pull of Mars's gravity. These solutions now predicted that the spacecraft would fly by the planet at an altitude of 110 kilometers—a distance that was very much on the low side, though still probably survivable. Nothing could be done now except wait and pray.

Two groups of scientists and engineers had gathered at JPL and Lockheed Martin for the critical orbital insertion event. NASA's cable channel began broadcasting the event live. Probably only a small group of insomniac space buffs watched the live broadcast, but several other TV channels sent cameramen and reporters to tape the event. Thurman and seven other JPL team members worked at computer terminals in a glass-lined control room, while the media and other curious onlookers peered in through the glass. A similar scene took place at Lockheed Martin's mission support center in Denver, though the working group there was much larger and the media representatives fewer. The dress code was shirtsleeves and jeans at the science lab, ties and slacks at the contractor.

At 1:41 a.m. in "Earth receive time"—eleven minutes after the corresponding events at Mars—signals arrived indicating that the Mars Climate Orbiter had begun retracting and stowing its solar panels in preparation for the orbital insertion burn. The process took eight minutes. Then the spacecraft began turning 180 degrees so as to convert its main rocket engine into a retrorocket. This took six minutes. At this point the spacecraft was heading over Mars's north pole, and NASA-TV showed an animation of the gracefully gyrating spacecraft as it cruised over the white expanses of frozen carbon dioxide that marked the pole. At 1:56, pyros (explosive devices) fired to pressurize the fuel and oxidizer tanks. As planned, the spacecraft stopped transmitting data: The only signal still being received at JPL was the single-frequency "carrier" signal. The tension was visible in the faces of the eight men in the control room.

Then at 2:01 a.m. came the voice of Lockheed Martin systems engineer Kelly Irish: "Real-time Doppler indicates main engine burn." In other words, engineers had seen that the frequency of the carrier signal was beginning to rise as the spacecraft decelerated under the influence of the retrorocket. It was a moment of exuberant relief in the JPL control room.

The group could follow the slowing of the spacecraft for the first four minutes of the planned sixteen-minute burn, but at 2:04 and 56 seconds the spacecraft's carrier signal began to break up, and six seconds later it disappeared completely. This was the expected effect of the spacecraft passing behind the planet into its radio shadow. "At this time we are in our occultation period," announced Irish.

The spacecraft actually went into occultation fifty-two seconds before the event was expected. Less than a minute early—surely that tiny error could be of no significance after a nine-month voyage? Irish's voice didn't betray any surprise, and the NASA-TV commentators continued their chatter about the details of aerobraking. But to Thurman, it was a bad omen. He started staring intensely at a sheet of paper that he was holding in his left hand, then glancing up at the screen in front of him.

"I remember I had a plot next to me that our mission engineer had made that allowed us to correlate the time that the spacecraft headed behind the planet. He'd come up with a clever scheme where, by looking at the time of loss of signal, we could get a guesstimate of the actual altitude. The lower the altitude, the earlier the loss of signal. I had this very quick way of watching data on the screen, seeing the loss-of-signal time, and then looking at the data sheet for the guesstimate. I remember we had the signal—it was very early, and I remember thinking, Uh-oh, this is not good. That really— Yeah, my anxiety level shot up, I tried to hold my composure since there were seven guys with cameras in front of me."

The NASA-TV commentator eventually seemed to pick up on the early occultation, because without mentioning the fact that it was early he attempted to explain it away. "The signal can be refracted by the atmosphere," he said. "The accuracy is not as deterministic as we'd like." Meanwhile, the eight men in the glass booth sat or stood there, fidgeting, looking at one another's screens, well aware that they could do nothing but wait twenty-one minutes for the scheduled reemergence of the spacecraft from occultation. By that time the spacecraft should have terminated its burn and turned so that its signals could be picked up by the 70-meter Tidbinbilla deep space antenna near Canberra, Australia.

At 2:26 a.m., the predicted time for the spacecraft to reemerge from occultation, there was dead silence in the control room, and even the television commentator stopped talking. The silence went on for minutes, while the men in the glass booth became increasingly fidgety, staring at their watches or at the computer screens, standing up and sitting down again for no apparent reason, or glancing into one another's tired, anxious faces. Sam Thurman kept adjusting his wedding ring, as if its precise alignment was crucial for the spacecraft's destiny. "Waiting for acquisition of signal," said the disembodied voice of Kelly Irish. And a minute later: "Still haven't seen anything; stand by."

After a few minutes Thurman stood up and began to talk with the other members of the JPL team, including the Mars '98 project manager, Richard Cook. His words weren't audible on the TV broadcast, so I asked him whether he had been telling them that the spacecraft was probably lost. "I don't say things like that," he said. "That can be devastating to a team's morale. You always have to hope for the best and do everything you can to make it come about. And there's a little bit of lore called the twenty-four-hour rule—which is, when something seemingly bad or scary happens, don't overreact for a minimum of twenty-four hours, because more often than not what actually happened and what you need to do about it might be different from what you think at first. So I tried to reach inside and gather the presence to say— Getting emotional doesn't serve any purpose, so you have to say, 'Here are the options: the vehicle survived and may be in safe mode; it might be tumbling; it might have ended up in a much lower or higher orbit than expected because of an overburn or an underburn.' Remember, the ground stations need fairly accurate predictions of what the transmitter frequency is going to be in order to tune their receivers properly to hear a spacecraft 120 million miles away, orbiting another planet. So it could be the spacecraft was there and transmitting fine, and the set of predicts we used to drive the antennas were off."

About thirty minutes later Richard Cook came out of the glass booth and spoke with NASA-TV. He outlined the possible factors that might have caused the spacecraft to go into safe mode. "At this point, we're still very confident that we're in orbit at Mars," he said, "and we're going to see the spacecraft signal sometime in the next few hours." With that, NASA wrapped up its live television coverage for the night.

Very quickly, however, devastating news came in. During the occultation period the navigators had been working on a revised prediction of the spacecraft's periapse altitude, based on data received shortly before orbital insertion. The results weren't good. "They came back in and said, 'Oh my God, this thing was sixty or seventy kilometers lower than we thought it was going to be,' said Thurman. "And that's when we thought, 'Oh boy, if it went that deep, it must have fried.'"

Engineers continued searching for a signal from the Mars Climate Orbiter for forty-eight hours, but it was largely a formality. Navigational errors, it now seemed clear, had led to the spacecraft dipping too deep into the Martian atmosphere during orbital insertion. The exact fate of the spacecraft was a matter for speculation. It may have broken up or exploded, scattering debris over the Martian surface—in which case there is some concern that terrestrial germs may have survived the heat of the reentry and contaminated the surface. Alternatively, frictional heating may have caused the retrorocket to cut out prematurely—in which case the spacecraft may not have been captured into orbit around the planet at all. In that case it would have continued forever on its lonely solar orbit, perhaps to be seen again by some distant generation of Earthlings or Martians.

JPL's John Casani was quickly appointed to head an internal investigation of the mishap. There was considerable time pressure, because the Orbiter's companion spacecraft, Mars Polar Lander, was fast approaching the planet and was due to arrive on December 3. There were many similarities between the two spacecraft, and investigators wanted to ensure that whatever doomed the Orbiter would not also affect the Lander.

All the focus was on the navigational problems. Every piece of navigational software was scrutinized line by line, and on September 29 an engineer identified the crucial error: the lack of a conversion factor to change pounds of force to newtons in the Small Forces software. Once that error was identified, all the problems with navigation were readily explained.

To my knowledge, Casani's report only circulated internally and was never published. "In my subjective opinion it focused too much on all the technical details of what didn't get done, or didn't get done well enough," commented Thurman, "and too little on how the lab got itself in that position to begin with." That deficiency was quickly made up for by a second investigation, headed by Art Stephenson, Director of NASA's Marshall Space Flight Center in Alabama. While agreeing with Casani that the units problem was the root cause of the mishap, Stephenson's report put far more emphasis on the numerous contributing factors—inadequate training, testing, and communication; the failure to resolve the anomalous navigational solutions or to report the problem through the proper channels; the failure to execute TCM-5; and so on—that together amounted to a systems-engineering or even programmatic failure. While not exactly criticizing Goldin's "Faster, Better, Cheaper" philosophy, Stephenson urged that it be practiced under a set of guidelines that he summarized as "Mission Success First."

Like the Mars Climate Orbiter, the Polar Lander was plagued with problems during its journey to Mars, but navigational errors were not among them; the navigational software did not contain the units error that destroyed the Orbiter. By the time the Lander reached Mars, it looked like all the kinks had been ironed out, but after it began its descent through the atmosphere, nothing more was heard from it. Once again, Richard Cook was delegated to give the waiting press an upbeat assessment of the situation. "I'm very confident the Lander survived the descent," he said, five hours after the Lander should have been broadcasting its first images. The reason for the Lander's demise was never established with certainty, but the most probable cause was a set of faulty sensors in the spacecraft's landing legs. These, investigators believe, told the Lander's computer that the spacecraft had landed while in fact it was still airborne. This caused the Lander to switch off its retrorocket prematurely and fall to the surface at a speed sufficient to destroy the entire spacecraft. Again, then, the root cause of the loss was attributed to a failure at Lockheed Martin, the manufacturer of both spacecraft.

The loss of the Lander, coming so quickly after that of the Orbiter, triggered an investigation that was independent of NASA. This one was headed by Tom Young, former executive vice president of Lockheed Martin—an odd choice, perhaps, given that anyone associated with the company might be expected to lay blame somewhere outside the company's purview. In fact, Young's report laid most emphasis for the failures on niggardly federal funding for the missions: They were underfunded by at least 30 percent, Young wrote.

Casani and Thurman concurred. "That was the problem," said Casani. "The only way that you could get the costs down was with less people." "Faster, Better, Cheaper was raging like influenza through the agency," said Thurman. "NASA and JPL should not have attempted to do two missions with that ambition and with that kind of cost and schedule. To me that's the fundamental root cause."

As the boss of NASA and a political appointee, Dan Goldin took a different line, accusing Lockheed Martin of having underbid to win the contract for the missions. "I think in this circumstance that the Lockheed Martin team was overly aggressive, because their focus was on the winning," he said in a PBS interview. "The Lockheed Martin Company did not pay attention, and I know it sounds like a paradox, but it was more important to them to win for today, and they didn't think of the long-term future or the reputation of their company."

In any event, Lockheed Martin Astronautics went through a tough period after the Mars '98 failures. The company suffered financial losses, NASA canceled a contract for a follow-up mission, and many astronautical engineers left the company, at least temporarily. But it rebounded, and in 2001 Lockheed Martin was awarded the contract for a successor to Mars Global Surveyor—a spacecraft named Mars Reconnaissance Orbiter. This was launched in 2005 and successfully went into orbit around Mars the following year. Orbital insertion was aided by a new technology, in which photographs of Mars's two moons were used to get a precise fix on the spacecraft's position as it approached the planet. As if to celebrate this success, in September of 2006 Lockheed Martin was awarded a multibillion-dollar contract, this time to build *Orion,* the successor to the Space Shuttle.

As to the mantras of "Faster, Better, Cheaper" and "Mission Success First," Steve Jolly expressed some skepticism. "I think what we do is avoid using any branding like that," he said. "Not because it's not fashionable anymore, but because slogans can sometimes hurt you. Now the real approach is, what's the right design to accomplish the objectives that we're being asked to do, and what's the do-able cost associated with that? What we're finding is that we can still leverage all those technologies and approaches that were developed in the nineties to pull off what we call 'best-value missions' for the government."

Although Lockheed Martin lost some employees after the Mars '98 failure, the hapless young engineer who made the units error was not fired; in fact, he's still with the company. "He has a lead position; he's in the critical path for all the flying missions that we have," said Jolly. "You know, that's the noble way to do it. Engineers do not walk in and say, 'I'm going to make a mistake today.'"

The Mars Climate Orbiter is the only U.S. space mission to have failed on account of a confusion of units, but several others have failed or gone seriously wrong on account of similarly "dumb errors" in data handling. In April 1999, for example, a *Titan* IVB rocket carrying a military satellite failed to reach orbit after its upper stage lost stability and broke up. The mishap was caused by the misplacement of a single decimal point in the control-system software.

Confusion of units has been the cause of many mishaps in other fields of science. Medicine and medical research has been particularly susceptible, most commonly with respect to drug dosages. A tragic example occurred in Ottawa, Canada, in 2002. Researchers at the Children's Hospital of Eastern Ontario were testing the use of the interleukin-2, an immune system booster, in the treatment of a childhood cancer known as neuroblastoma. The first patient, one-year-old Ryan Carroll, became severely ill during the drug treatment, but he survived. Rather than halting the trial, the researchers proceeded to treat another patient, four-year-old Ryan Lucio, with the same drug regimen. After four injections he suffered multiple organ failure and died. At this point, Ryan Lucio's doctors went back and examined his treatment plan, and they discovered a terrible mistake. Instead of calculating the dose of interleukin-2 in units of micrograms per square meter of body area as they should have done, they had calculated it in micrograms *per kilogram of body weight,* which meant that they had given him an approximately twenty-five-fold overdose. They soon realized that Ryan Carroll had been overdosed in the same way.

The U.S. Food and Drug Administration, which had approved the trial, went ballistic and posted an excoriating critique of the trial's principal investigator, Dr. Jacqueline Halton, on its Web site. Health

Canada, on the other hand, expressed little, if any, criticism and quickly issued its approval for the trial to continue. It may be that Health Canada's mild-mannered approach reflected guilt that it had allowed the two Ryans to be experimented on at a time when the proper application and safety assurances had not been provided by the researchers, contrary to Canadian law.

CHAPTER 11

※

SPEECH PATHOLOGY:
The Monster Study

On the morning of January 17, 1939, a twenty-three-year-old graduate student named Mary Tudor set out with five colleagues from the campus of the University of Iowa at Iowa City. Their destination was the Iowa Soldiers and Sailors Orphans' Home in the town of Davenport, fifty miles to the east on the Mississippi River. Tudor's mission was to discover the cause of stuttering. She didn't accomplish her mission, but the methods she used in her attempt to do so are now— sixty-eight years later—the subject of a multimillion-dollar lawsuit and the focus of fierce ethical controversy.

The experiment that Tudor planned to perform on the orphanage children was directed by her advisor, Wendell Johnson, a thirty-two-year-old assistant professor in the Departments of Psychology, Speech, and Child Welfare at the University of Iowa. Johnson, who died in 1965, devoted his life to the study of stuttering. He was largely responsible for transforming stuttering from a risible handicap to a topic of serious academic and clinical research, and many of the leading experts in the field today are his students or "grand-students." Thus, although the particular theories he espoused may not have stood the test of time, he is revered as a founder of the field and a longtime advocate for people with speech disorders.

Johnson himself stuttered. In a 1930 book, he described how he began to stutter at the age of five, after several years of normal childhood speech. Nothing in particular seems to have provoked the onset of the disability—no physical illness, personal loss, or traumatic experience. Nor were there other stutterers in Johnson's family. In fact, Johnson's childhood was quite typical for a boy born into a rural Kansas household in the early 1900s.

During his first few years of stuttering the trait caused him relatively little hardship. He was held back a year in school on account of his disability, but he was not punished for it or subjected to any unusually harsh remedies. For the most part, his family and his few playmates tolerated his stuttering amicably; they would sometimes help him out by completing the words that he stumbled on. In fact, his stuttering helped motivate him to shine in ways that compensated for the disability—in academics, sports, and in jovial social interactions. Thus, paradoxically, the experience of being a person who stuttered may have been a positive factor in Johnson's career and in his personal life.

At the age of sixteen Johnson was subjected to the first serious effort to eliminate his stutter: He was sent for the summer to a residential school that offered a program for this purpose. Johnson and his fellow students were taught to speak in a slow, drawling monotone. They were taught to speak rhythmically, while swinging Indian clubs or doing other exercises to set the rhythm. They were taught to ignore other people's negative reactions to their stuttering. And they tried to follow the director's exhortation to "Use your will power. Don't give up. Be the master of your fate and the captain of your soul!"

And to a point, it worked. Johnson became less afraid of stuttering, and perhaps as a consequence he stuttered less or not at all—within the protective environs of the school. As soon as he reentered the wider world, however, his stutter returned with full ferocity.

After completing high school with honors—he was class valedictorian—Johnson attended nearby McPherson College. While there, he learned that a program of research on stuttering was being started at the University of Iowa's Speech Clinic, and he transferred to that university in 1926. Johnson recounts how, as one of his first tasks there, he had to read aloud for five minutes before a class of students: During that time he was able to get four words out of his mouth. In spite of that inauspicious start, Johnson remained at the University of Iowa for his entire life. He obtained a B.A. in 1928 and a Ph.D. in psychology in 1931, and was appointed assistant professor in 1937 and full professor in 1945.

During his early years at the university, Johnson's advisors suspected that stuttering resulted from a developmental miswiring of the brain—specifically, an error in the assignment of functions to the left and right hemispheres of the cerebral cortex. They came up with the idea that Johnson, who to all appearances was right-handed, would be better off left-handed. For several years, therefore, Johnson was equipped with a variety of devices that prevented him from using his right hand and forced him to use his left. Other stuttering students were similarly outfitted, and they became a familiar sight on campus, their good arms bandaged or tied back as they struggled to perform life's tasks with their clumsier arms.

This and a variety of other experiments failed to cure Johnson or his fellow students of their stutters. And gradually Johnson began to reject the general theory that stuttering was caused by some in-born miswiring of the brain, and to consider a very different set of ideas based on the premise that social interactions were the key to the disorder.

Most central to Johnson's new thinking was the notion that the very act of labeling a child a stutterer might turn him or her into one. In part, this idea was influenced by Johnson's reading of the work of the Polish-American psychologist Alfred Korzybski, founder of a field he called General Semantics. Korzybski was interested in the impact of labels on people's perceptions of things. According to one anecdote, for example, he shared some cookies with his students, and after they had eaten them showed them the package, which read DOG COOKIES. This caused some of the students, who had previously enjoyed the cookies, to rush to the toilet to vomit. Interestingly, this demonstration brought up—in miniature—some of the same ethical concerns that plagued the Tudor study.

But why would anyone label a child a stutterer if he or she did not already stutter? According to Johnson, it was because *all* young children mangle their speech to a certain extent. These normal childhood "disfluencies" are typically ignored by parents, teachers, and peers, and they disappear over time as the child's speaking skills improve. Some parents, however, develop an inordinate concern with their children's disfluencies, Johnson believed. They become obsessed with the notion that the child is beginning to stutter, and they communicate that obsession to the child, calling him or her a stutterer and drawing unnecessary and repeated attention to every inconsequential error of speech that the child may commit. In doing so, the parents do not merely impose on the child the identity of a stutterer, but they also inculcate the child with an intense fear of stuttering. Thus the child, in speaking, anticipates making errors, becomes increasingly tense, and fights his or her own vocal organs. This inner battle leads to the syllable repetitions, prolongations, and other phenomena that are the behavioral hallmarks of stuttering. Johnson's ideas became known as the "diagnosogenic" theory of stuttering, so called because the trait originates in the very act of diagnosis.

Was Johnson's diagnosogenic theory the trigger for the Tudor study? Not according to Ehud Yairi, Professor Emeritus in the Department of Speech and Hearing Science at the University of Illinois at Urbana-Champaign. Yairi is an expert on stuttering who entered the University of Iowa as a graduate student shortly after Wendell Johnson's death. In a 2005 article, he concluded that the diagnosogenic theory could not have been well enough established in Johnson's mind to have been the basis for the 1939 study, because Johnson's first published account of it appeared in 1942, three years after the Tudor study was completed. At the time of the study, Yairi argued, Johnson was still thinking in terms of neurological explanations for the disorder.

Yet other experts have reached different conclusions. One of these is Nicoline Ambrose, an associate professor in the same department as Yairi, who collaborated with him on a detailed reanalysis of the Tudor study published in 2002. When I asked Ambrose in a 2006 interview whether she thought that the Tudor study was intended as a test of Johnson's diagnosogenic theory, she said, "I would basically say yes—or some earlier version of it as it was being formulated—although I don't believe the intent was to create stutterers, but to invoke stuttering on a temporary basis. I don't think there was any intent to say, 'Let's see if we can create a long-term problem in these kids.'" Another expert who has weighed in with a similar opinion is Oliver Bloodstein, a onetime student of Johnson who is now Professor Emeritus of speech at the City University of New York. Bloodstein has written that Johnson was already entertaining the central idea of the diagnosogenic theory in the years prior to Mary Tudor's study, and it was indeed this theory that led him to initiate the study.

I recently stumbled on a little-known lecture by Johnson that he published in 1938—one year before the Tudor study—under the title "The Role of Evaluation in Stuttering Behavior." This lecture laid out the core of the diagnosogenic theory and even claimed to provide evidence in support of it. "[I]n 92 percent of the 47 child stutterers we have studied to date," he wrote, "the first order reaction was a simple, loose repetition of sound, syllable or word. When this was negatively evaluated—disapproved by the parents and then by the child—other reactions appeared in series. The higher order reactions tended to be more complex, involved more tension, and more stoppages generally." In other words, castigating a child for run-of-the-mill disfluencies caused them to spiral into full-scale stuttering.

What's more, the internal evidence of the Tudor study itself strongly implies that it was designed as a test of the theory. Tudor's master's thesis, which was based entirely on the study, was titled "An Experimental Study of the Effect of Evaluative Labelling on Speech Fluency." This is the Introduction section, in its three-sentence entirety:

> Certain published statements (Johnson, *Language and Speech Hygiene*) and examination of case histories suggest the possibility of regarding the diagnosis of stuttering as one of the factors responsible for the development of the disorder.
>
> An investigation of the effects, particularly on speech fluency, of such a diagnosis is indicated from this point of view. In view of this consideration the present study has been done.

In other words, the *only* reason that Tudor put forward for undertaking the study was to test the diagnosogenic theory of stuttering.

In the second section of the thesis, titled Problem, Tudor stated that the study was designed to answer the following questions:

1. Will removing the label "stutterer" from those who have been so labeled have any effect on their speech fluency?
2. Will endorsement of the label "stutterer" previously applied to an individual have any effect on his speech fluency?
3. Will endorsement of the label "normal speaker" previously applied to an individual have any effect on his speech fluency?
4. Will labeling a person, previously regarded as a normal speaker, a "stutterer" have any effect on his speech fluency?

Evidently, the study was intended to test the effect of evaluative labeling—specifically, labeling as a stutterer or a normal speaker—on children's speech. Although the written objectives do not spell out what the resulting "effects on speech fluency" might be, it's reasonable to assume that they were expected to consist of the appearance or disappearance of stuttering—either of the complete phenomenon or some of its components—at least on a temporary basis, for otherwise the study does not make a great deal of sense. The use of the more general phrase "effects on speech fluency" may have reflected the experimenters' open minds about what the results of the study might be. More likely, though, it represented a wish not to spell out too baldly one of the study's ethically troubling goals: the attempt to elicit in normal children the very trait that had plagued Wendell Johnson for his entire life.

Looking more closely at the objectives, one can see that objectives 2 and 3 are, by themselves, pointless. No one would expect that continuing to use labels that have been previously applied to a person would have any interesting effect on their speech fluency. Evidently, these two "objectives" were not really intellectual goals in themselves but were listed merely as a way of indicating the need for control groups—subjects who were not manipulated and who were therefore not expected to show any effects. It is objectives 1 and 4, which involved changing a child's previously applied labels, that incorporated the real goals of the study.

The children at the orphanage—a mix of real orphans and children whose parents had been forced to give them up by economic necessity—were used to being treated as guinea pigs. When Jim Dyer, a reporter, interviewed one of the now-elderly subjects for a 2001 article in the *San Jose Mercury News,* she told him that "Every week somebody else from the university would come down and start testing us for God knows what." There is no record of Wendell Johnson's motive for choosing an orphanage for the study, but we may guess that it was twofold: First, the convenience of having a large, fairly homogeneous collection of children at a single location and cared for by the same staff, and second, the ease of obtaining permission for the study. It would have been much harder, one may guess, to get permission from a child's parents, given that one possible outcome of the treatment was the development of a speech disorder.

The first day of Mary Tudor's visit to the orphanage was devoted to the selection of children for the study. There were ten so-called stutterers in the orphanage—children who the teachers and matrons considered to be stutterers and had labeled as such. All ten of these children were included in the study. To balance them, Tudor and her five colleagues—fellow graduate students who were familiar with speech disorders—picked twelve children at random from the remaining population of children who had never been called stutterers by the staff. The twenty-two children selected for the study included both boys and girls, and their ages ranged from five to sixteen years.

Each of these two groups was then further divided into two, thus providing the four subject groups needed for testing the four objectives described above. Tudor named the groups IA, IB, IIA, and IIB, but for ease of recall I'll rename them as follows:

S→N: Five children previously labeled as stutterers who were to be relabeled as normal speakers.

S→S: Five children previously labeled as stutterers who would continue to be labeled as such.

N→S: Six children previously labeled as normal speakers who were to be relabeled as stutterers.

N→N: Six children previously labeled as normal speakers who would continue to be labeled as such.

In her thesis Tudor maintained her subjects' confidentiality, only referring to the individual children by code numbers. This confidentiality was breached by the *Mercury News* reporter Jim Dyer, however. The names of the six children in the N→S group (the most ethically questionable group, consisting of normal-speaking children who were to be relabeled as stutterers) have also entered the public domain on account of the lawsuit against the state of Iowa in which they or their heirs are plaintiffs, and I will therefore use their names here. They were Norma Jean Pugh (age five at the time), Elizabeth Ostert (nine), Clarence Fifer (eleven), Mary Korlaske (twelve), Phillip Spieker (twelve), and Hazel Potter (fifteen).

The ages of these children immediately raise a significant issue with regard to the scientific value of the study. Stuttering typically begins in the preschool years; if Wendell Johnson began stuttering at five, as he related, then he was among a minority of late-onset stutterers. The children in Mary Tudor's N→S group, with the possible exception of Norma Jean Pugh, were well beyond the age at which stuttering typically develops. Thus even if Johnson's diagnosogenic theory were correct, Tudor's study might have failed to validate it— simply because the children had grown past the sensitive period of speech development during which they could be induced to stutter. Tudor did not discuss this issue in her thesis. It may be that she was forced to use older children because there was an insufficient number of younger ones in the orphanage. Alternatively, she may have felt compelled to use children in the same age range as those in the stuttering groups, who averaged twelve years of age.

The plan of the study was as follows. At the beginning and again the end of the study, the speech of all twenty-two children was to be assessed by the panel of five judges. Without knowledge of which experimental group each child belonged to, the judges would independently provide a numerical assessment of the child's fluency and would also make a judgment as to whether the child stuttered or not. During the intervening four months Tudor would apply labels to the children according to the groups they had been assigned to.

This is how Tudor's thesis describes what was to be said to the children in the N→S group at the beginning of the study. (Her actual words were modified to suit each child's age and intelligence; some of the children had IQs that were well below average.)

> The staff has come to the conclusion that you have a great deal of trouble with your speech. The type of interruptions which you have are very undesirable. These interruptions indicate stuttering. You have many of the symptoms of a child who is beginning to stutter. You must try to stop yourself immediately. Use your will power. Make up your mind that you are going to speak without a single interruption. It's absolutely necessary that you do this. Do anything to keep from stuttering. Try very hard to speak fluently and evenly. If you have an interruption, stop and begin over. Take a deep breath whenever you feel you are going to stutter. Don't ever speak unless you can do it right. You see how [the name of a child in the institution who stuttered rather severely] stutters, don't you? Well, he undoubtedly started this very same way you are starting. Watch your speech every minute and try to do something to improve it. Whatever you do, speak fluently and avoid any interruptions whatsoever in your speech.

The children in the S→N group were told the opposite—that they didn't stutter, that any speech mistakes they made were inconsequential and that they should not worry about them. The children in the S→S and N→N groups were given messages consistent with their prior identities as stutterers or normal speakers respectively.

Tudor reinforced these messages on subsequent visits to the orphanage. She had eight or nine sessions with each of the children in the N→S group, and three or four sessions with the children in the

S→N group. The thesis doesn't mention any sessions with the children in the S→S or N→N groups: Either she neglected to list these sessions, or perhaps she thought that their status as controls made the sessions unnecessary.

During the sessions with the children who were being relabeled as stutterers, Tudor would pick on slight speech errors that the children made in the course of their conversation and drew attention to them, saying that they were signs of stuttering and that the child should do everything in his or her power to avoid making the errors. In addition, Tudor attempted to recruit the orphanage's staff to help reinforce these messages. She told them that the N→S and S→S children were stuttering and that they should draw the children's attention to all their speech errors. Similarly, she told the staff that the S→N and N→N children were not stutterers, and she asked them to ignore these children's speech errors or to tell them that their speech was fine.

It seems that the staff didn't cooperate in the fashion that Tudor hoped. Although a couple of the children in the N→S group mentioned to Tudor that their teachers had commented on their speech, Tudor wrote in her thesis that the staff generally didn't follow her instructions, or only did so to a small degree. Thus the overall amount of indoctrination that the children received was probably much less than Tudor originally desired.

Even so, the indoctrination clearly had an effect. Here is part of Tudor's report of an interview with one of the children in the N→S group, eleven-year-old Clarence Fifer, on May 2—three and a half months into the study:

> "How is your stuttering today?"
> "I don't know."
> "When do you seem to have the most trouble?"
> "When I'm playin'."
> "Tell me something about it."
> "Well, most of the time I stutter."
> "Do the other boys notice it?"
> "Sometimes."

"Do they ever say anything?"

"No."

"How do you know they notice it?"

"They kinda laughed."

"What did you do then?"

"Walked away."

"Does it bother you much?"

"Yes, feel pretty bad."

"What do you do about it?"

"Next time try to keep myself from doin' it."

"How do you do that?"

"Sometimes I take a breath."

"How does it feel when you speak?"

"Kinda strain my throat."

His speech had a breathy quality and he took a breath after every few words whether he needed it or not.

During this interview he had twenty-five speech interruptions. The stuttering phenomena added to the previous list were deep inhalation, excessive exhalation, and eyes closed.

Since Tudor deceived the staff in the same way that she deceived the children, the staff could not have been in a position to give any kind of informed consent to the study. Whether there was any person at the orphanage, such as its administrator, who was informed about the true purpose of the study is not stated in Tudor's thesis. Jim Dyer, who interviewed Tudor in 2000, when she was eighty-four years old, wrote that Johnson obtained permission for the study from orphanage officials, but he didn't make clear whether Johnson actually told these officials what would be done to the children. It's possible that Johnson felt he had carte blanche to initiate any kind of study that he considered appropriate.

So what was the result of the study? What happened to the children's speech? In his articles in the *San Jose Mercury News* in 2001,

Dyer reported that most or all of the children in the N→S group responded to being labeled as stutterers by—stuttering. In doing so, Dyer said, they confirmed Wendell Johnson's diagnosogenic theory. In addition to stuttering, Dyer reported that many of these children became withdrawn and isolated; they were reluctant to speak at all, and what few words they did speak came out in single words or brief phrases rather than complete sentences.

When Dyer tracked down some of the children—now elderly adults—for his articles, they supposedly confirmed the findings of the study. Norma Jean Pugh, who was five at the time of the study and spoke normally, apparently told Dyer that she had been induced to stutter by Tudor's experiment, and that her stutter persisted for years, gravely damaging her social relationships and her education. Now, at age sixty-four, she was a near-total recluse. Mary Korlaske, who also spoke normally at the start of the experiment, was also induced to stutter, she told Dyer. She later got over the stuttering, but it recurred in 1999 after the death of her husband. She moved into a retirement home, where she rarely left her room. Dyer said that she stuttered when he interviewed her, although his description of her speech did not correspond closely to what a speech pathologist would call stuttering.

There is some evidence that Johnson too believed that labeling the children as stutterers caused at least some of them to stutter. In e-mail correspondence, Johnson's student Oliver Bloodstein told me that "In his lectures in the fall of 1942, Johnson made it clear that he thought the results of the Tudor study supported the diagnosogenic theory." Bloodstein also wrote (in a published article): "To the best of my recollection, he told us that one child actually did begin to stutter as a result of the procedure."

Although Dyer visited the University of Iowa library, where Tudor's thesis is archived, he did not say explicitly that he read the thesis, and most of his account is based on interviews with Tudor and the surviving subjects, along with readings of Tudor's notes. (I was not able to locate Dyer for an interview.)

A totally different account of the Johnson-Tudor study was published in 2002 by Nicoline Ambrose and Ehud Yairi, the experts on stuttering at the University of Illinois. Ambrose and Yairi actually went back and read the sixty-year-old typescript that was Tudor's thesis, and what they wrote about it in the *American Journal of Speech-Language Pathology* contradicted the central assertion of Dyer's articles: Tudor's experiment, they said, did not cause any of the children to stutter.

This conclusion was based principally on the assessments of the children's speech that were made at the beginning and end of the study by the panel of five blinded judges. Each judge independently rated the fluency of the children's speech on a five-point scale, with 1 corresponding to the worst fluency and 5 to the best. At the beginning of the study, the average score for the children in the crucial N→S group was 2.83—roughly in the middle of the scale of fluency rather than near 5 as one might expect. At the end of the study the average score for these children was 2.92. Statistically, the tiny shift of the average (by 0.09 units) was completely insignificant, and what's more, it was a shift toward improved speech—the opposite of what Johnson's theory would have predicted. The child in this group who showed the biggest shift was Mary Korlaske, who supposedly told Dyer that she was induced to stutter by the experiment. Her speech shifted by 0.8 units—in the direction of greater fluency!

The fluency scores given by the judges included sub-scores for individual kinds of disfluency, some of which (such as repetition of syllables) were symptoms of stuttering, while others were not. Even when Ambrose and Yairi looked specifically at the scores for the stuttering-related disfluencies, there was no significant change over the course of the study.

Besides the numerical score, the judges added written comments at the end of the study. For each of the five children, including Korlaske, the majority of the judges simply wrote "no stuttering." Some of the judgments included statements like "appeared hesitant" or "answered briefly," but not one judge stated that any child stuttered or mentioned repetition of syllables, the key symptom of stuttering.

None of the other three groups showed any significant shift in their average speech fluency either. Even looked at individually, none of the children showed any substantial shift in the direction predicted by the theory. Thus, Ambrose and Yairi's analysis showed that Dyer's central claim—that the treatment caused the normally speaking children to stutter—was wrong. Nor, apparently, had the stuttering children been caused to stop stuttering by being labeled as normal speakers.

One might think on this basis that the Tudor study was actually a refutation of Johnson's theory rather than a confirmation, since changing the children's labels had no effect on their propensity to stutter. But no; it was worse than that, according to Ambrose and Yairi. They reported that the study was so poorly designed and executed that it could not have been expected to reveal anything about the theory, regardless of whether the theory was right or wrong.

Most crucially, Ambrose and Yairi reported that many of the children had been assigned to the wrong subject groups. If they had been correctly assigned, the children in the N→S and N→N groups should have been given scores near the 5 (fluent) end of the scale, and the children in the S→N and S→S groups should have been given scores near the 1 (disfluent) end of the scale. In fact, however, there were no significant differences between the average scores of any of the groups before the treatment began. There were several children who were clearly described as stuttering or repeating syllables who were put in one of the "N" groups, and several children who were described as not stuttering who were put in one of the "S" groups. Apparently, the "stuttering" children were selected simply because the orphanage staff said that they stuttered, and the "normal-speaking" children were selected simply because the staff said that they didn't stutter, and even though these assignments were not always borne out by the judges' assessments, the children were left in the groups they were assigned to.

Thanks to an interlibrary loan, I was finally able to lay hands on Tudor's thesis myself, and I confirmed the truth of what Ambrose and Yairi said on these points. However, this problem is not as devastating to the credibility of the study as Ambrose and Yairi implied. For one thing, many of the items used for fluency scoring were not criteria used in the diagnosis of stuttering, and for these unrelated items there was no reason to expect that stuttering children should score differently from nonstuttering children. Also, Tudor was concerned with the effects of changing *labels:* thus in selecting children for the study what mattered most was how a child was *labeled* prior to the study, not whether he or she actually stuttered or not. In this sense, Tudor had good reason to depend on the judgments of the orphanage staff, who had been in contact with the children for years.

There were other reasons, however, why no solid conclusions could be drawn from the study. I already mentioned the fact that the children were too old to test the diagnosogenic theory if children's susceptibility to criticism was limited to a developmental period around the age when children typically begin to stutter. Also, the children formed an unrepresentative sample in many respects, such as being institutionalized and also in most cases having below-average IQs. Furthermore, there were too few children in each group for it to be likely that significant effects of treatment would emerge.

Finally, the indoctrination of the children was done ineffectually. As already mentioned, the staff didn't cooperate, leaving Tudor's visits as almost the only "relabeling" that the children experienced. It is hard to believe that just a few sessions with Tudor would somehow outweigh a lifetime of being exposed to the opposite labels. Tudor herself commented on this in the Discussion section of her thesis: "As it was," she wrote, "the children received their stimulation almost entirely from the writer. If these children had been constantly reminded of their speech they would have undoubtedly reacted more positively [i.e., by showing more signs of stuttering]." And she predicted that "more extensive results" could be expected if the experiment had been done in a "home situation" with constant critiques from the children's parents.

Although, in the Results section, Tudor reported the onset of "stuttering phenomena" during the treatment of some of the N→S children, such as Clarence Fifer, she did not mention any induction of stuttering in her Discussion. This was presumably because neither the judges' assessments nor her own numerical analyses of the children's speech documented such an effect. But she did conclude that these children were affected in other ways:

All of the subjects in Group IIA [N→S] showed similar types of speech behavior during the experimental period. A decrease in verbal output was characteristic of all six subjects; that is, they were reluctant to speak and spoke only when they were urged to. Second, their rate of speaking was decreased. They spoke more slowly and with greater exactness. They had a tendency to weigh each word before they said it. Third, the length of response was shortened. The two younger subjects responded with only one word whenever possible. Fourth, they all became more self-conscious. They appeared shy and embarrassed in many situations. Fifth, they accepted the fact that there was something definitely wrong with their speech. Sixth, every subject reacted to his speech interruptions in some manner. Some hung their heads, others gasped and covered their mouths with their hands; others laughed with embarrassment. In every case the children's behavior changed noticeably.

This description is similar to Dyer's assertion that the children in the N→S group became withdrawn, isolated, and reluctant to speak. The notion that Tudor perceived that the children suffered this kind of harm is bolstered by the fact that, after completion of the study she returned to the orphanage three times to "debrief" the children in the N→S group—that is, to reassure them that they were in fact normal speakers and to encourage them to speak freely.

After the first of these visits, in March 1940, she wrote to Johnson as follows: "I didn't find them as free from the effects of the therapy I had inflicted upon them last year as I had hoped to. But as I am still a firm believer in the theory of evaluative labeling, I wasn't too disappointed." This passage (quoted by Dyer) confirms that Tudor perceived that the children suffered some psychological harm from the study, and that this harm lasted at least through to the following year. In addition, it implies that she saw this as confirmation of Johnson's theory. Either she was suggesting that the children were in fact stuttering, or she thought that any kind of difficulty in speaking, even a reluctance to speak motivated by low self-esteem, would count as a confirmation.

In their analysis, Ambrose and Yairi took a cautious view of the idea that the N→S children suffered lasting harm. "There did seem to be an effect that when children were told 'You'd better watch out what you say' or 'You'd better not repeat,' they got anxious about talking," Ambrose told me. But she pointed out that there was no objective assessment of such effects in the study. In addition, she tended to pooh-pooh the notion that the children had been turned into asocial hermits by the experience, as some of the elderly survivors apparently claimed in their discussions with Dyer. "These were children who were given up [by their parents], children who were living in a situation way less than ideal," Ambrose said. "These individuals had so many difficulties in their lives, I don't see why you would think that that particular study of a few months, with very little contact really, had any kind of significant effect on their lives." She mentioned an episode recounted by Dyer, in which Mary Korlaske ran away from the orphanage and made it all the way back to her mother's home, whereupon her mother summoned the police to take Mary back to the orphanage. "I would think that would be a much more problematic situation than someone telling you that you should speak fluently," Ambrose commented.

Given that Ambrose and Yairi are members of the academic speech-pathology community, it may be that they were motivated to minimize the harm suffered by children in Tudor's experiment as a way of protecting Johnson's reputation. Nevertheless, they minced no words in their critique of the Tudor study's scientific merits. All in all, it seems very difficult at this point to discern what role, if any, the Johnson-Tudor study played in the difficulties that some of the N→S children experienced in their later lives.

Tudor's thesis was never published, and Johnson never referred to it in his own published writings. The only reference he made to the study was in his lectures to students, in which he cited it in support of the diagnosogenic theory, as mentioned by Oliver Bloodstein. According to Jim Dyer, this failure to publicize the study amounted to a positive "hiding" of it. Dyer maintained that the reason had to do with the atrocities perpetrated by Nazi doctors in the name of medical science during World War II. After those crimes came to light, the entire scientific community was sensitized to ethical issues in research involving human subjects, and in this new environment Tudor's experiments would have seemed ethically indefensible.

Ehud Yairi took exception to Dyer's imputation that Johnson actively hid the Tudor study. Yairi stressed that the thesis has been available in the University of Iowa library and that at least nineteen people are recorded as having read it between 1941 and 1993. Yet shelving an unpublished master's thesis in a university library is equivalent, in the majority of cases, to destroying it. Such theses are rarely read and almost never cited.

Given that the study, on its face, was actually a *disconfirmation* of his theory, Johnson's failure to publish it or refer to it in print could be viewed in quite a different light—that is, as part of a strategy to protect his pet theory from findings that put it in doubt. If so, his and Tudor's failure to publish would be ethically troubling. It is equally possible, however, that Johnson realized that the study had grave methodological shortcomings that would not have passed scrutiny. "In light of today's standards, I would say this is totally not publishable because it's a very poorly done study," commented Ambrose.

Although Tudor's thesis languished on the library stacks, and Tudor herself left the University of Iowa to take up a position as a speech therapist, there was some institutional memory of her experiment. Some of Johnson's students began referring to it as the Monster Study. As far as I know the first use of this phrase in print was in 1988, when the late Franklin Silverman, a professor of speech pathology at Marquette University, used it as the title of an article about the study.

During the postwar years Wendell Johnson gradually built up his diagnosogenic theory into the centerpiece of his thinking about how stuttering developed and how it should be prevented or cured. It was but one of many "blame the parents" theories that were current in that epoch: Schizophrenia, autism, homosexuality and many other traits were said by influential academics and doctors to be caused by the way parents and others treated young children.

Although Johnson held fast to his theory until his death, other lines of research gradually made it seem less and less plausible. For one thing, studies of young children—some of them done by Nicoline Ambrose—revealed that the speech mistakes made by children at the very onset of stuttering are not the normal disfluencies exhibited by other children, but distinct abnormalities that are immediately recognizable as the beginning of stuttering. A child can be talking entirely normally, Ambrose told me, and the next day he or she wakes up obviously stuttering. There is increasing evidence that there is a genetic predisposition to stuttering. The efficacy of some drugs in relieving stuttering also points toward a biological explanation and away from a theory based on family dynamics. At this point, Johnson's diagnosogenic theory is dead, and most of the current research interest is in pinning down the brain miswiring that was hypothesized by Johnson's teachers in the 1930s.

That is not to say that Johnson's contributions were worthless. He made useful studies of the "stuttering block"—the mental events that lead to the moment when the person who stutters does actually stutter. And Johnson is still considered to have greatly helped people to overcome or reduce their stuttering by encouraging them to see the trait as something they could control—an echo of the "Be the master of your fate and the captain of your soul" mantra that he first heard as a teenager.

Ambrose emphasized this positive aspect of Johnson's work in her discussion with me, but she also added a warning that it sometimes leads to negative consequences, especially a tendency to blame the person who stutters for his or her failure to improve. "[Johnson's] idea that if stuttering can be learned it can be unlearned has done a disservice in some ways," she said, "because if you keep on stuttering, what's wrong? And we don't think that's a failure on the part of the clinician or the client, if they're not able to reduce their stuttering."

During all the years in which Johnson's diagnosogenic theory gained credence and then gradually lost it, the Tudor study remained largely unknown. Tudor herself, as well as the subjects of her study, led inconspicuous lives. There were occasional written references to the study, such as the 1988 article by Franklin Silverman, mentioned earlier. And in 1999 Jerome Halvorson—a onetime professor of speech pathology at the University of Wisconsin who'd obtained a master's degree at the University of Iowa—wrote a novel about the study, titled *Abandoned: Now Stutter My Orphan*. Halvorson used pseudonyms for the students rather than divulging their real names. Because the novel was self-published and highly idiosyncratic in style, it attracted little attention.

It was Jim Dyer's articles in the *San Jose Mercury News* in 2001 that first drew wide attention to the Tudor study. Dyer's articles contained what purported to be accurate details of the study, interlaced with sentimental accounts of a human drama that was largely orchestrated by Dyer himself.

A brief account of how this happened was given to the *Des Moines Register* by David Yarnold, executive editor of the *Mercury News*. Yarnold said that Dyer, besides his job at the *Mercury News*, was also a graduate student at the University of Iowa. Using his identity as a graduate student, Dyer gained access to the Iowa State Archives in Des Moines—specifically to confidential records that are only open to academics for bona fide research purposes. From these records, Yarnold said, Dyer obtained the real names of the children in Tudor's study. (However, some of these names were also mentioned in Tudor's notes, according to Dyer.) Armed with the names, Dyer tracked down and interviewed several of the surviving children—now elderly adults. He told them about the real purpose of Tudor's study, which the children had apparently never been informed of, even in the course of Tudor's "debriefing" sessions after the study was completed. (Again, I haven't been able to confirm this account with Dyer himself.)

Several of the subjects reacted to the information with understandable anger. Dyer describes the lives of some of them as having followed a downward spiral, starting with the stuttering that was allegedly caused by the study and progressing to near-complete social isolation in some cases. The centerpiece of Dyer's story was Mary Korlaske, now a widow and a reclusive inhabitant of a retirement home, who was so incensed by what she heard from Dyer that she wrote an angry letter to Tudor—whether spontaneously or at Dyer's suggestion, I don't know. The letter, which was addressed to "Mary Tudor the Monster" concluded as follows:

> As I sit here crying . . . I wondered what I could say or send you to remind you of the hurtful pain that never goes away.
>
> I'm sending you your own thimble.
>
> God try to have mercy, or should he? You had no mercy for the children who still cry in the night.
>
> —Mary Korlaske Nixon Case No. 15
>
> P.S. When the tears get realy bad, punch a whole in the bottom of the thimble like I did. Then the thimble won't over flow.

Dyer hand-carried the letter to the eighty-four-year-old Tudor—or else, he just happened to be present when it arrived in the mail—and he described Tudor's reactions as she looked it over: the shaking of her head, the trembling of her hands, and her comment, "Oh dear—I hope it isn't a bomb."

According to Dyer, Tudor herself made both positive and negative comments about the study. On the one hand she was proud of a study that—as she still believed—proved Johnson's diagnosogenic theory correct. "It was a small price to pay for science," Dyer quoted her as saying. "Look at the countless number of children it helped." On the other hand, she expressed shame at the apparent harm the study had done to some of the children. "That was the pitiful part—that I got them to trust me and then I did this horrible thing to them," she said. Tudor deflected much of the blame onto Johnson who, she said, told her to perform the study in the fashion that she did, and neglected to have psychotherapists help the children to recover from the trauma that had been inflicted on them. I had hoped to hear more about this from Tudor herself, but when I tried calling her in the fall of 2006 a neighbor informed me that she had died a few weeks earlier.

Dyer resigned from the *Mercury News* after the ruse he had allegedly used to discover the names of Tudor's subjects came to light, and since that time he seems to have disappeared from public view.

In a lawsuit brought against the State of Iowa in 2003, the three surviving subjects in the N→S group, along with the estates of the three who had died, requested unspecified damages for intentional infliction of emotional distress, fraudulent misrepresentation, breach of fiduciary duty, invasion of privacy, and civil conspiracy. I asked Curtis Krull, the attorney representing Mary Korlaske in the case, whether it was Korlaske's contention that she had been made to stutter, as Dyer alleged. Krull said that was not part of the allegation; rather, the claim was that Korlaske was put into a state of believing that she had a speech impediment, and that this caused lifelong psychological problems such as insecurity and low self-esteem. The State of Iowa attempted to have the case thrown out on the basis of its supposed immunity to prosecution under the law as it existed in 1939, but in 2005 the Iowa Supreme Court ruled in the plaintiffs' favor, and the case is now back at the trial-court level and proceeding through the discovery process. Often, cases such as this one end up being settled out of court.

As suggested earlier, it may be very difficult to tease apart the harm caused by the Tudor study from that caused by the many other traumas suffered by the orphanage children both before and after they participated in the study. But this uncertainty has not prevented many people from passing judgment on the ethical issues surrounding the case.

One person who has reviewed these issues in particular detail is Richard Schwartz, Presidential Professor in Speech and Hearing Sciences at the Graduate Center of the City University of New York. Schwartz has for many years chaired CUNY's Institutional Review

Board. In 2005 Schwartz published an analysis of the Tudor study with this question in mind: Would the study be approved if it were proposed to an IRB today? Schwartz believes that there are several reasons why it would not. First, there was the real possibility for harm to the children who were to be labeled as stutterers. "If you really believe the theory, you're going to turn these children into stutterers," he told me in a 2006 interview. In addition, there was the possibility for more general psychological harm, as the lawsuit alleges did occur. These potential harms were not balanced by any potential benefits to the children.

Second, Schwartz says that the study would be judged unethical on account of its poor design and execution, which made it unlikely that it would add anything to the fund of knowledge about stuttering. If a study cannot generate useful findings it is unethical to engage human subjects in it.

Third, Schwartz believes that the study would be judged unethical because of its use of institutionalized children, who are considered to lack the same protections and capacity for choice that children living with their parents typically enjoy. In fact, current federal regulations would rule out the Tudor study on these grounds alone.

Lastly, Schwartz believes that the use of deception in the Tudor study would be considered unethical today, because the scientific issues at the heart of the study could have been addressed by other means, and because the deception was not justified by any probability that real advances in scientific understanding or human welfare would result from its use.

Schwartz confessed to me that (like so many other people who have taken an interest in the case) he had not actually read the Tudor study but was dependent on information and extracts provided by others, especially by Ambrose and Yairi.

Most commentators have expressed criticisms of the Tudor study similar to those put forward by Schwartz. Ambrose and Yairi, for example, wrote that "It is unquestionable that the study was ethically wrong." But one person has mounted a vigorous defense of Wendell Johnson—his son. Nicholas Johnson, a law professor at the University of Iowa, wrote an article it which he maintained that historical research should be judged only by the ethical standards of its own time. He then came up with a laundry list of equally or even more questionable studies from that general period, including the infamous Tuskegee syphilis study in which poor black men were denied access to treatment for their syphilis for many years. According to the younger Johnson, his father and his father's student did nothing that was outside the bounds of normal research practice at the time.

Much of what Nicholas Johnson says is perfectly true: Researchers did often take advantage of institutionalized persons for research in those days, sometimes inflicting worse harm on them than what Mary Tudor probably did to her subjects. But the interest in revisiting ethical questions about a historical study such as Johnson's is not—except perhaps for Johnson's son and a bunch of lawyers—to pass retroactive moral judgment on the deceased persons who conceived it and carried it out. Rather, it is to highlight the reasons why it is necessary to have written regulations governing the use of human subjects in research today, as well as IRBs to enforce them.

Nicholas Johnson has also attempted to defend his father by pointing out that Tudor herself—who has largely escaped personal criticism—should share whatever blame is assigned for the study. Certainly graduate students need to accept responsibility for their actions, but in this particular case it is clear that the project was entirely Wendell Johnson's idea. When I read Tudor's thesis, I was struck by her near-total lack of interest in the issues that her project addressed; even though her Introduction briefly mentioned Johnson's diagnosogenic theory as the inspiration for the study, her Discussion—also very brief—included no assessment of what her findings meant for the theory. In general, Tudor's thesis reads like the work of an industrious low-level operative who followed her advisor's instructions to the letter, and who considered her job complete when she had done so.

"Whether there was true harm or not, [Johnson and Tudor's] subjects were intruded on in a way that they shouldn't have been," commented Schwartz by way of wrapping up the ethical issues. "They should have given this more thought, even given the mores of the time. Most importantly, it's a useful thing today to teach both senior researchers like myself, and students, that you really have to think about these things. It's very important, if anything, to err on the side of being cautious and more protective of human subjects and to be really good at perspective-taking: 'What would this be like if this were my child, my relative, me, in the situation of being a subject; would this be OK?' And this is really at the heart of what an IRB tries to do."

CHAPTER 12

⊗

NUCLEAR CHEMISTRY:
The Magic Island

In August 1999, nuclear chemists at the Lawrence Berkeley National Laboratory announced the creation of three atoms of a new "super-heavy" element, element 118. Two years later they had to retract their claim, and a firefight broke out that cost a star scientist his career and sullied the reputations of several others.

The University of California's Berkeley campus had been a world leader in the discovery of new elements since 1940, when Edwin McMillan discovered element 93, neptunium. The university's most famous element hunter was Glenn Seaborg, who discovered plutonium (element 94) in 1941 and followed it up with nine more elements, culminating in 1974 with the one that was named in his honor—seaborgium (element 106). McMillan and Seaborg shared the 1951 Nobel Prize for Chemistry, but many other Berkeley scientists also played important roles in this work. These included Stanley Thompson, who helped discover most of the new elements up to element 101, and Albert Ghiorso, who shared credit for many of the elements discovered from the mid-1940s onward.

The use of the term "discover" in this context is slightly odd. Darleane Hoffman (also a Berkeley Lab scientist since 1984) and others did discover minute amounts of plutonium and neptunium in natural uranium ores, but none of the other "transuranium" elements exist in the natural world, unless perhaps in some distant supernova. Thus the process of discovery means *creating* them, not finding them as the term implies. In part, scientists use the term *discover* simply as a continuation of a tradition that started with the actual discovery of the lighter elements in nature. In addition, however, they probably use the term because they think of the transuranium elements as already existing in a Platonic universe to which their powerful instruments give them entry. They think this way because the properties of each element—even those that don't exist—were fixed at the beginning of time, when the particles that make up atomic nuclei (the positively charged protons and uncharged neutrons) were endowed with their immutable characteristics.

To some extent, then, nuclear chemists can predict the properties of atomic nuclei that haven't yet been discovered or created. The most basic theoretical formulation is this: The protons and neutrons are held tightly together by the "nuclear force," which only acts over minute distances. Countering this attraction is the electrostatic repulsion between the protons' positive charges, which tends to push them apart; this force acts over a much greater distance than the nuclear force. As one progresses to heavier and heavier nuclei, they become less and less stable, because the protons and neutrons cannot crowd closely enough together for the nuclear force to act at full strength between all the particles. Thus the electrostatic repulsion comes to dominate, causing the nucleus to break apart.

If this were the whole story, there would have to be an end to the periodic table of the elements, and that end would lie somewhere in the neighborhood of element 106, the last element that Seaborg discovered. During the 1980s and 1990s, however, elements 107 to 112 were created—roughly in the sequence of their atomic numbers. Most of these discoveries were made by a group at the Institute for Heavy Ion Research (GSI) in Darmstadt, Germany. A Russian group, the Joint Institute for Nuclear Research in Dubna, near Moscow, was also a player.

The existence of these heavier elements had in fact been predicted by theorists who went beyond the simple model of the atomic nucleus just described. One of these theorists was Seaborg's Polish-born colleague Wladyslaw Swiatecki, who joined the Berkeley group in 1957. Swiatecki and others believed that, within the nucleus, protons occupy a series of discrete energy levels that can be thought of as concentric shells. A nucleus whose outermost proton shell was completely filled would gain an extra measure of stability beyond that predicted by classical theory. Similarly, neutrons were thought to reside in their own shells and to confer extra stability on the nucleus when their outermost shell was filled. These nuclear shells are analogous to the better-known electron shells outside of the nucleus, which are filled in the inert gas elements helium, neon, and so on.

The numbers of protons and neutrons that conferred stability were said to be "magic," and a nucleus that contained magic numbers of both protons and neutrons were "doubly magic." These might exist in sizes far beyond the limits set by classical theory. In other words, even element 112 might not be the end of the road.

Not everyone agreed on exactly what these magic numbers of protons and neutrons were, or even whether they were meaningful concepts at all. Still, this was the conceptual framework that guided research in the 1990s. And what it meant was that simply going for the next-heaviest undiscovered element on the list might not be the best approach: Some elements well beyond the presently achieved limits might actually be more stable and easier to create.

Also, this approach meant that both the number of protons (which defines which *element* we are talking about) and the number of neutrons (which defines which *isotope* of that element we are talking about) needed to be considered when thinking about creating superheavy elements. To illustrate this, Seaborg, Swiatecki, and others used a chart that plotted the proton number (on the vertical axis) and the neutron number (on the horizontal axis) of all known atomic nuclei. On this chart, the already-known nuclei formed a long, narrow cluster running from the bottom left (a hydrogen nucleus) toward the top right (the currently heaviest element). The cluster resembled New York's Long Island as seen on a map. Outside this cluster lay a "sea of instability" in which nuclei could not exist, or not for long enough to be detected.

Yet across this sea in a direction farther upward and to the right (corresponding, say, to the locations of Block Island and perhaps Martha's Vineyard), might lie "islands of stability" or "magic islands"—the homes of yet-undiscovered superheavy nuclei with sets of protons and neutrons close to doubly magic numbers. The rumored existence of these islands offered as powerful a lure to nuclear chemists as the fabled Spice Islands did to the explorers of old.

There seem to have been some differences of opinion within the Berkeley Lab concerning these ideas. Darleane Hoffman and Albert Ghiorso clearly believed in the idea of islands of stability or magic islands, because in a 2000 book they frequently used these phrases when describing their laboratory's goals. I got a different story from Walter Loveland, a somewhat younger nuclear chemist from Oregon State University who joined the Berkeley Lab for the 1998–1999 year, and who played an important role in the ill-fated search for element 118. "I would disabuse you of this idea of the 'island of stability,'" he said in a 2006 interview. "Those predictions were made in the 1960s when it was thought that there would be a group of elements with half-lives that were long even relative to the age of the universe, and that they'd form this island. We don't believe in that anymore—that's not right. What we know is that there may be nuclei whose half-lives are longer than their neighbors', but they seem to be connected to the mainland of lighter elements by a peninsula. They are not islands in a sea of instability."

By 1998, when the effort to detect element 118 began, Berkeley's glory days of element hunting were long over. Stanley Thompson had died in 1976.[*] Seaborg was eighty-six, and in August of 1998 he suffered a devastating stroke that led to his death six months later. Ghiorso was eighty-three; Swiatecki and Hoffman, the youngsters, were seventy-two and seventy-one. Room 307 of Berkeley's Gilman Hall, where Seaborg identified plutonium, had been a National Historic Landmark for thirty-two years. And though much other good work had been done at the Berkeley Lab, not a single new element had been discovered there in more than two decades.

The Berkeley Lab did have the tools to produce superheavy

[*] Ghiorsi called Thompson on his deathbed to announce the lab's discovery of a new superheavy element, element 116. Thompson died a few hours later, but the "discovery" later turned out to be an error.

elements, however. One essential tool was the 88-inch Cyclotron—the giant descendant of the hand-held "proton merry-go-round" invented by Ernest Lawrence in the 1920s. The Cyclotron accelerates nuclei of a chosen isotope (let's say ^{48}Ca, which are calcium nuclei with 20 protons and 28 neutrons, yielding a total mass number of 48) to speeds that can exceed 1,000 kilometers per second, giving them tremendous kinetic energy.

A steady beam of these energetic nuclei emerges from the Cyclotron and enters another piece of equipment, the gas-filled separator. This instrument was built by a Berkeley group led by Ken Gregorich, who belongs to a younger generation; he was one of Seaborg's last graduate students. Within the separator, the beam passes through a thin foil made from another isotope, such as ^{244}Pu (plutonium with 94 protons and 150 neutrons). The hope is that a very occasional beam nucleus will strike a target nucleus just right. If so, the kinetic energy of the beam nucleus overcomes the electrostatic repulsion between the two negatively charged nuclei (known as the Coulomb barrier) and the two nuclei fuse, yielding a compound nucleus. In this example it has 114 protons (making it the as-yet-unnamed superheavy element 114) plus 178 neutrons.

Because of the kinetic energy of the incoming ^{48}Ca missile, the compound nucleus is put in an excited state, like a drop of water brought nearly to a boil. Most of this excitation energy is carried off almost instantly by the "evaporation" of a few neutrons. The remaining, slightly lighter isotope of element 114 flies onward through the instrument, and a series of magnets deflect it from the main beam and deposit it onto a silicon detector. This nucleus is itself unstable: It breaks down into lighter nuclei over some period of time that might range from microseconds to minutes. The detector identifies the time, location, and energy of the particles produced in these sequential breakdown events, and from this data the superheavy nucleus that gave rise to them can be identified. Voilà, a new element—element 114 in the case of this hypothetical example—albeit just a single atom and a very short-lived one at that.

That was the basic idea, but up until then it hadn't worked. Either the energy of the incoming nucleus was too low to overcome the reticence posed by the target nucleus's Coulomb barrier, or it was too high and the target nucleus simply disintegrated, like a toad struck by a flying princess. There didn't seem to be any "just-right" level of energy—the gentle kiss that would allow the long-hidden Prince Charming to step forth.

Things seemed to change in 1998. A young Polish theoretical physicist by the name of Robert Smolańczuk, who was then at the Berkeley Lab's German rival, GSI, did new calculations of the expected probabilities for the creation of superheavy elements by various fusion reactions. He reported that if atoms of ^{208}Pb (the commonest natural isotope of lead) were bombarded with a beam of ^{86}Kr nuclei (an isotope of the inert gas krypton) that had been accelerated to exactly the right energy, there was a surprisingly good probability of scoring hits that would generate nuclei of element 118—a superheavy element far beyond what had been created up until then. "What Smolańczuk picked up on," Gregorich told me, "was that you need to bombard at an energy level that's just over the Coulomb barrier: This happens to be the correct energy needed to make the product."

Nuclear chemists calculate the probability of such successful hits in units called barns, whose name derives not from some famous scientist named Barn or Barnovsky, but from the phrase "can't hit the broad side of a barn." Most theorists thought that the probability for such reactions fell off exponentially with increasing atomic number, so that by the time one got to element 118, it might be down in the femtobarn range (one quadrillionth of a barn) or less, making the reaction essentially unachievable even in a lifetime of trying. Smolańczuk, on the other hand, pegged the probability as being many orders of magnitude higher, at 670 picobarns—nearly one billionth of a barn. This was still a challenging proposition but within the realm of possibility.

According to Smolańczuk's calculations, the compound nucleus formed by the fusion reaction would evaporate just one neutron. The reaction would be a "cold fusion," contrasted with the "hot fusion" reactions in which the compound nucleus was more highly excited and evaporated several neutrons, like the ^{48}Ca/^{244}Pu example described earlier. Thus the reaction would leave an atom of $^{293}118$—a nucleus with 118 protons and 175 neutrons. On the basis of shell theory, this atom seemed to lie on the western shore of a magic island: it could probably exist for a very brief period—a fraction of a millisecond—but it would have too few neutrons to last for longer.

Darleane Hoffman, still a leader of the Berkeley group in spite of having been officially retired for seven years, invited Smolańczuks to join the lab, which he did in October of 1998. His ideas got a very enthusiastic response. Hoffman, along with Albert Ghiorso, urged the junior members of the team to put Smolańczuk's reaction to the test, and quickly. Smolańczuk had told the German and Russian groups about his ideas, and so the quest for element 118 had suddenly become an international horse race—in Hoffman and Ghiorso's minds, at least.

Ken Gregorich, the designer of the gas-filled detector, and Walter Loveland, the visitor from Oregon State, took the lead in setting up the experimental apparatus. For the data analysis, they turned to Victor Ninov.

Ninov was born in Bulgaria, but had obtained his doctorate at GSI before coming to Berkeley. While at the German lab he took part in the research that led to the discovery of elements 107 through 112. Darleane Hoffman and Ken Gregorich hired Ninov away from GSI in 1996. Because of his achievements at the German lab, the move was thought to be quite a coup for the Berkeley group—an acquisition that would greatly increase their chances of finding a superheavy element of their own.

At Berkeley, Ninov worked closely with Gregorich on the construction of the gas-filled separator and its associated instruments. His most valuable sphere of expertise, however, was in developing and using the software that analyzed the output of the detectors. This software had to search the instruments' output files, which were binary files recorded on magnetic tape. The software looked for events whose time, location, and magnitude corresponded to what would be expected for the breakdown of the nuclei that were being sought.

In the case of element 118, the original $^{293}118$ nucleus was expected to decay by giving off a sequence of alpha particles, each of which consisted of two protons and two neutrons. These alpha emissions were what the detector actually detected. The first alpha emission would mark the breakdown from $^{293}118$ to $^{289}116$—itself an undiscovered element. The next would mark the breakdown of that nucleus to $^{285}114$, and so on all the way down to $^{269}106$, which is seaborgium. The entire cascade was expected to take about two seconds. The software was designed to pick this characteristic chain of alpha emissions out of millions of irrelevant events.

Ninov was clearly under a great deal of pressure in the last few months of 1996. "We . . . convinced Victor Ninov that the reaction should be run as soon as possible," wrote Ghiorso and Hoffman later, "as we greatly feared GSI or Dubna might do it first." Whether this pressure came only from Ghiorso and Hoffman, or also from Gregorich and Loveland is not clear. When I talked to the latter two men in 2006, both of them suggested that the experiment was planned more as a shakedown cruise for the new equipment than as a confident attempt to find a new chemical element.

For several months, technical difficulties with the apparatus delayed them. The general level of motivation jumped considerably in January of 1999, however, when a report came in that the Dubna lab had created a single atom of element 114, using the calcium-plutonium reaction described earlier.

The Russians had run through more than a million dollars' worth of ^{48}Ca to achieve that one seemingly successful strike. If superheavy nuclei are defined as those with an atomic number greater than 112—the usage favored by Hoffman and Ghiorso—then this was the first superheavy nucleus ever created. What's more, the nucleus stayed intact for all of half a minute before breaking down. If this finding was genuine, the Russians had already landed on or near an island of stability. "[W]e felt happy that at last the Magic Island had been found," wrote Ghiorso and Hoffman in their 2000 book, "and we redoubled our efforts to get our own experiment under way."

On April 8, 1999, the Berkeley experiment finally began. Over a period of four days, the lead foil target was bombarded with 700,000,000,000,000,000 krypton nuclei, each of them boosted to an energy of nearly half a million electron volts. When Ninov applied his software to the resulting data, he found two alpha-decay chains. The energies of the alpha emissions and the time intervals between them were remarkably close to the values predicted by Robert Smolańczuk. It seemed clear that the run had produced two atoms of element 118, which had decayed into another never-before-seen element—element 116—and then into element 114 and even lighter elements.

To be sure of the result, the group ran another experiment a couple of weeks later, in which they hurled more than twice as many krypton atoms at the target as they had done during the first run. The researchers were therefore expecting that they might see four or more alpha chains. In fact they only saw one, but it was a beauty, again confirming Smolańczuk's predictions. So the research team assumed that the lower yield on the second run was just a statistical fluctuation, and they added the one sighting to the previous two, meaning that three atoms of element 118 had now been created.

"Does Robert talk to God or what?" exclaimed Ninov, according to Ghiorso and Hoffman's memoir. There was general amazement at the close fit between the observations and Smolańczuk's theory. Clearly there was some initial worry about this among the researchers, but the worry eventually gave way to jubilation. "It was such a startling discovery that strenuous efforts were made to find out if anything had gone wrong," wrote Ghiorso and Hoffman, "but nothing obvious was uncovered . . . Now there is no question, the Super-Heavy Island actually exists!"

The findings were quickly written up and submitted to *Physical Review Letters,* a journal that specializes in rapid publication of newsworthy findings. The paper appeared in print on August 9, 1999, only three months after the experiments were completed. The paper had fifteen authors. First came Ninov, Gregorich, and Loveland—the central players—followed by a group of other faculty members who had played at least a peripheral role, including Ghiorso, Hoffman, and Swiatecki. The list was rounded off with a gaggle of graduate students who, as Loveland put it, "took shifts minding the separator in the middle of the night and stuff like that."

The paper described the observation of the three alpha decay chains and the evidence that these had originated in three nuclei of the new element 118. From their data, the authors calculated that the probability of production of element 118 by the krypton-lead fusion reaction was about 2 picobarns—well shy of Smolańczuk's estimate of 670 picobarns, but still an amazingly efficient reaction compared with what most nuclear chemists would have predicted.

The publication of the paper was accompanied by excited pronouncements from the scientists involved. "We jumped over a sea of instability onto an island of stability that theories have been predicting since the 1970s," said Ninov, according to the *Berkeley Lab Research Review*. In June, Hoffman and Ghiorso added a bubbly epilogue and some more exclamation marks to their already overloaded book. "We have convincing evidence for elements 114, 116, and 118!!" they wrote. They mentioned their sadness that Seaborg had not lived to witness the discovery. (Seaborg was a posthumous coauthor of the book, however.) All the major U.S. newspapers carried stories about the discovery, often on their front pages. It was an American flag, after all, that the Berkeley group had planted on the Magic Island.

Very quickly, rival laboratories geared up to duplicate the feat of Ninov and his colleagues—but they couldn't. Over a period of a year or so, GSI, GANIL (a French laboratory), and later RIKEN (in Japan) all announced their failure to create even a single 118 nucleus by the krypton-lead reaction, even with levels of bombardment that made the Berkeley Lab's efforts seem like a mild peppering.

Meanwhile, back in Berkeley, the researchers were tearing down and rebuilding their equipment, including the gas-filled separator, in preparation for new studies. It was a time of great optimism. "It was clear that once this reaction worked, there were many other reactions you could open up—you could almost discover the other chemical elements one by one in a straightforward manner," Loveland told me. Loveland himself wrote a whole new suite of analytical software in preparation for the new work.

The Berkeley group's reaction to the negative reports from overseas was fairly dismissive; they believed that the equipment in those other laboratories didn't have the requisite sensitivity to replicate their own findings. In March of 2000, Darleane Hoffman was awarded the American Chemical Society's Priestley Medal. In her acceptance lecture she gave pride of place to the discovery of element 118.

Still, the Berkeley group eventually realized that it might be a good idea to repeat their own work before venturing further. So in early 2000 they conducted two more runs, totaling about the same length as the successful runs of the previous year. Not a single element 118 nucleus was spotted—a statistically very improbable result, assuming that the earlier runs were authentic. The group put together a committee of nuclear science specialists from other Berkeley labs to examine the problem, and in January of 2001 that committee wrote a report that focused on possible problems with the equipment during the 2000 runs, as well as on ways to resolve them. It seemed likely that Gregorich and his colleagues just needed to retune everything and they would soon get back to their winning ways.

Following the unsuccessful 2000 runs the group did in fact spend about a year checking and improving their equipment. The next run began in April of 2001. This run, in Loveland's phrase, was "massive": The beam, the detectors, and many other aspects of the experiment had been so thoroughly upgraded that the researchers expected to reap a rich harvest of element 118 atoms. Yet several days went by with no signal. Then, finally, Ninov announced success: He had found just one alpha decay chain with the unambiguous signature of an element 118 nucleus.

"Victor came up with the chain," said Loveland, "and Don Peterson—who was the postdoc working with me—and myself were there, and we said, 'Victor, we want to be able to see that chain,' because we now had our own software to analyze the data. We said, 'Let's go ahead, tell us where it is on the tapes, we'll look at it.' We tried; it was an eventful weekend. We tried very, very hard, and had absolutely no success. We couldn't find this event at all. . . . We talked to Victor and said, 'There's a terrible problem, we can't find this.' He said, 'Oh, I'll show you,' and he pulled it up on his computer screen, and I said, 'Oh my God, this is really tough, because now we're caught in a situation where it depends whose software we're using, and there's something terribly wrong at that point.'"

So far, it looked as if the problem was a technical one involving the data-analysis software. Either Ninov's software was finding events that didn't exist, or Loveland's was failing to find events that *did* exist. A second investigatory committee was formed in June of 2001, headed by Darleane Hoffman. The committee delegated a postdoc and Victor Ninov to independently search for the reported decay chains in the output files from both the 1999 and the 2001 runs: Neither of them could find any of the chains. Meanwhile, Ken Gregorich got into the act. He wrote an entire data-analysis program from scratch in an attempt to resolve the conflict. When he applied this program to the data, he again failed to see any sign of the four decay chains that Ninov had reported.

Because the problem now seemed to be in the data-analysis software rather than the equipment, a third committee was assembled, this time consisting of Berkeley computer experts, including Gerald Lynch, Augusto Macchiavelli, and two others. At some point in this process, some of the magnetic tapes that carried the original output from the detectors went missing, and they have remained missing ever since. The Lynch-Macchiavelli group also found that some of the crucial files from Ninov's own computer were missing. Loveland had copied all of Ninov's directories onto his own computer, however, and the investigators relied in part on those copies in their further studies.

One thing that greatly aided the computer sleuths in their investigations was the existence of a "log file" in the data-analysis software. This file was a record of all the activities that had taken place during the process of analysis, as well as who had performed them. The log file showed that Ninov had indeed run an analysis on the data from the 2001 run and that it had come up with the element 118 decay chain, just as he had told his colleagues. However, another similar analysis run later with the same software had not shown any such decay chain. The sleuths went through these two files, checking every tiny detail, and they found evidence that the first file had been manipulated—some of the details such as the timing of the supposed events just didn't make sense, and the lengths of the pages were not what would have been generated during normal operation. They then showed that the apparent record of a decay chain in this file could easily have been inserted by cutting and pasting code from elsewhere in the software. When the investigators turned their attention to the records from the 1999 runs, they found similar evidence of manipulation. In all cases the suspect manipulations had been made from Victor Ninov's computer account. Furthermore, in 1999 at least, Ninov was the only person in the group who knew how to run the data-analysis software.

Once they realized that the alpha decay chains were not in the data, Gregorich and Loveland hastened to let the scientific community know—Loveland mentioned it at a research conference in New Hampshire in late June of 2001, for example, and the Berkeley group also put out a news release on the topic at the end of July. They knew that it was also necessary to sent an official retraction to the journal in which the original report had appeared, and Gregorich sent off the retraction statement at the beginning of October. But Victor Ninov sent his own letter to the journal, in which he asked for his name to be removed from the retraction statement. This put the editors of *Physics Review Letters* in a quandary. They sat on the letter for a year, while correspondence went to and fro between the editors, Ninov, and the other authors. Finally in July of 2002 the journal announced that "all but one" of the authors had requested a retraction—they didn't name Ninov as the standout. That put the paper into a kind of permanent vegetative state in which it still lingers today, although everyone knows that the Berkeley Lab's claim to have discovered element 118 is dead.

As soon as Ninov's onetime collaborators in Germany heard of what had transpired, they were concerned that similar invalid data might have been used to document the discovery of some of the elements he had worked on there. So the GSI scientists went back and examined all thirty-four decay chains that had been used to document the discovery of elements 110, 111, and 112. They found that two of the chains—the second of the chains reported in the element 110 study of 1994, and the first of those reported in the element 112 study of 1996—were based on "spuriously created" data. Sigurd Hofmann, the team's leader, told the Berkeley group about this, and in due course the GSI group published a retraction of those specific chains. This did not put their claims for discovering elements 110 and 112 into doubt, however, because additional, genuine chains were observed in the same and subsequent experiments. Although the published retraction did not go into details, Sigurd Hofmann told colleagues and reporters that the spurious chains had been created intentionally—by Victor Ninov.

"That was the killer," said Walter Loveland. The world was now closing in on Ninov. The university convened a fourth and final investigative committee, headed by Rochus ("Robbie") Vogt, a retired Caltech physicist. The committee surveyed the results of all the prior investigations and interviewed the participants—except for Ninov, who declined to attend committee meetings after the first introductory session. In March 2002 the committee issued its report, which named Victor Ninov as a perpetrator of scientific fraud.

The other investigators took some heat too—not for faking data, but for failing to independently analyze the data before publication. "The Committee finds it incredible," Vogt wrote, "that not a single collaborator checked the validity of Ninov's conclusions. . . . The claim of an important discovery demanded no less."

Neither Gregorich nor Loveland was very happy with this part of the report. Gregorich told me that he first set eyes on the report when it was sent to him by a journalist, and he felt aggrieved that he hadn't been given the chance to "check it for accuracy." Loveland took issue with the general notion that scientists are obligated to validate the authenticity of their colleagues' work. "In theory the answer is, yes, we should do this," he said, "but in practice the way science is carried out in large collaborative experiments is that people have assigned responsibilities, and if they're reliable and there's no question about their work, that kind of double- and triple-checking doesn't occur. There's just not enough time for anyone to do that sort of thing, so you trust people to do various portions of the experiment."

Of course, the main focus was on Ninov. In November of 2001, even before the Vogt committee met, he was put on paid leave and banned from the Berkeley Lab. In May of the following year he was fired.

After Victor Ninov left Berkeley he took a teaching position at the University of the Pacific in Stockton, California. When I tried to reach him there in the fall of 2006, I was told that he had left the college and that no one knew how to contact him—something that turned out not to be quite true, as (unknown to me) Ninov's wife was still a professor at the college. Eventually I was able to track him down, and he agreed to a telephone interview, which took place late one evening in December of 2006.

I started the interview, as I usually do, by asking whether I could tape the conversation. Ninov said that I could not, because previous journalists had twisted his words; the only journalists who had treated him fairly, he said, were Germans. Casting about for something that might change his mind, I offered to conduct the interview in German if I could use the recorder. To my surprise he agreed, and our conversation proceeded in that language for a few minutes. Unfortunately, I soon found that my German—which has been languishing in a dark corner of my cerebellum since my student days in that country—wasn't up to the technicalities we were discussing, so I lapsed into English. Ninov quickly pounced. "Are you still taping the conversation?" he demanded. "Yes," I said. "So you have already broken your word," he replied, and he demanded that I stop the tape and erase what I had already recorded.

At that point I thought the interview was about to come to a premature end, but in fact Ninov kept talking for an hour longer. He talked fast, sometimes about highly technical matters, and he didn't always pause to deal with my requests for clarification. Thus I don't claim to have understood everything he said, and I certainly wasn't able to write much of it down. However, I did take home a number of key points.

Most importantly, Ninov emphatically rejected all the accusations that have been made against him. "I did not do anything wrong," he said. "I stand by the integrity of everything I've done in my scientific life." And he denied he was ever less than frank in his dealings with his colleagues. "I presented everything openly to everyone; I never tried to hide anything."

Ninov told me that he had never liked the Berkeley Lab. The intellectual atmosphere was stuck in the 1950s, he said, and he didn't trust anyone there except Gregorich. The equipment was outdated and the resources meager. In building the gas-filled separator, for example, he had been forced to recycle decades-old magnets that he found in a storeroom. The main thing that kept him at Berkeley was its coastal location: He kept an oceangoing sailboat in the Berkeley marina, and since leaving Berkeley he has sailed it across the Pacific. (The GSI, he said disdainfully, is surrounded by forest.) In addition, he married an American soon after his arrival.

He also told me that he had never wanted to participate in the search for element 118. "I resisted to the bitter end to do the experiment," he said. His reasons had partly to do with the fact that the gas-filled separator wasn't ready—it wasn't yet possible to accurately measure the number of krypton atoms impinging on the target, for example. This criticism is substantiated to some extent by what I read in Hoffman and Ghiorso's book: They spoke of modifications being made to the setup even after the runs got under way. They also wrote of having to put pressure on Ninov to do the experiment, as mentioned earlier.

He was also unenthusiastic about the experiment because of his skepticism about whether Robert Smolańczuk's reaction would work. He told me that he asked Smolańczuk how far off base his "670 picobarn" estimate might be, and Smolańczuk replied, "Three orders of magnitude" (a thousandfold); this could put the actual figure around 1 picobarn. Again, Ninov's statement is substantiated by others. "Everybody expected this [670 picobarn] estimate to be wrong," Loveland told me. "Robert himself thought it would be wrong."

What about the three alpha decay chains that he reported finding in the data? Ninov denied that he had ever referred to them as decay chains. He called them "events," and he said that they could equally well have been artifacts caused by "damped oscillations." A damped oscillation would be a series of waves in the electronic equipment that rapidly diminished in amplitude. He suggested that each wave might cause a signal resembling an alpha emission, and the decreasing heights of the waves would cause the appearance of decreasing alpha energies, as expected for a decay chain. Even if they were actual decay chains, he said, there was no way to tell whether they arose from the decay of element 118 nuclei, because it was impossible to answer the key question—how many protons they possessed.

As far as I could tell from the Vogt report, Ninov never brought up his damped-oscillation explanation during the investigations. I asked him how such phenomena could give rise to precisely the half-lives and energies that were predicted by Smolańczuk's theory. His answer was that the electronic filters were set to only accept the predicted events out of a huge sea of candidate events, so it wasn't surprising that a few would match the requirements. Ninov emphatically denied that he had ever said, "Does Robert talk to God or what?" as claimed by Hoffman and Ghiorso.

Of course, I asked Ninov why he had gone ahead and published the claim for element 118 if, in his own mind, the data was so suspect. This is where he really surprised me. He told me that Gregorich and Loveland had jumped to the conclusion that the "events" were alpha decay chains, and that they had submitted the paper to *Physics Review Letters* without his knowledge or consent. He himself wrote an entirely different paper, he said, but the other two men rejected it, citing his poor English. (In speaking with me, Ninov's English was flawless—which is remarkable, considering that it is probably his fourth language.)

These claims don't square very well with the quotation cited earlier from the Berkeley Lab's magazine, in which Ninov reportedly talked in enthusiastic terms about the findings at around the time the paper appeared. Nor do they square with Ninov's refusal to sign the retraction letter. If he had been as skeptical about the authenticity of the data as he now claims, he should have been the first to call for a retraction. I didn't ask Ninov about these apparent contradictions. When I asked Gregorich to comment on Ninov's version of events, he sent me three excerpts from Ninov's draft of the paper, all of which interpreted the findings as detections of element 118.

With regard to the investigations that led to his identification as the person who had inserted the evidence for decay chains into the computer files, Ninov denounced the committees as biased against him and politically motivated. His colleagues also betrayed him, he said, especially Gregorich. With regard to the Lynch-Macchiavelli committee, he expressed contempt. "You know already from their names what's up," he said, implying that they were a lynch mob and were following the cynical recommendations of Macchiavelli's fifteenth-century namesake. "The investigation was absolutely political," he said.

I asked Ninov about the allegations that he had created spurious data at GSI before he came to Berkeley. Ninov denied it, and suggested that the Germans' statements were made in response to pressure from Ken Gregorich, who needed something to bolster the believability of his own allegations. Ninov criticized the Germans for cutting off communications with him, and was particularly bitter that he was not invited to participate in the choice of a name for element 111, which he helped discover. (The final choice was roentgenium.)

The persecution continued after he was ousted from Berkeley, Ninov said. An anonymous caller, whom he suspected was Augusto Macchiavelli or one of his colleagues, called the administration at the University of the Pacific and complained about his presence there. As a result, Ninov said, his appointment there was not renewed and he had to seek work at a different institution. He would not tell me where he was currently working, because he feared further persecution of the same kind. He did tell me that he had put the Berkeley episode behind him, however. "I've just moved on with my life and left the little animals to do what they want," he said.

Although Ninov's Berkeley colleagues experienced direct injury to their reputations as a result of being associated with an apparently fraudulent study, their expressed reactions have been more of puzzlement than anger. "He had nothing to gain," said Loveland. "He was a very successful scientist." Quite a few people have brought up possible medical explanations—perhaps something related to a head injury that Ninov suffered years ago. "There were a lot of discussions on whether there was an issue of painkillers involved, or issues of mental illness," said Loveland. And he recounted something that seemed to strengthen such ideas in his own mind. "I remember sitting out on the patio of the Cyclotron and asking him, 'Hey, Victor, what's going on?' He spun off some comments to me about conspiracies, and I thought, Oh dear, we have a real problem here, when people start talking about conspiracy theories and so forth."

Both Ninov and his wife (in a brief conversation I had with her) strongly rejected such ideas. ("Painkillers?" retorted Ninov. "Aspirin, maybe.") Both of them interpreted these speculations as indications of malevolence on the part of the people who expressed them. To me, they seemed more like attempts on the part of Ninov's onetime colleagues to construct explanations that removed some of the blame from a person they had once liked and admired.

One idea that many writers have stressed when considering scientific fraud is that the perpetrator fabricates data to prove an idea that he or she is already convinced is correct: This way, the perpetrator is unlikely to be found out, because future genuine findings will confirm the fraudulent ones. The spurious data in the GSI studies could be taken in support of this idea. The first apparently fraudulent decay chain only cropped up after one genuine chain was detected. Thus Ninov, presuming that he was the perpetrator, might have thought that he was merely hurrying the process of discovery along a little by adding a spurious chain. Once that fraud passed muster, perhaps committing further frauds became easier—maybe even addictive.

One thing stuck in my mind that Ninov told me in passing. "I was never interested in nuclear chemistry," he said. At first, I thought that this was just a bitter comment on a career that had culminated in such ignominy. Then I wondered. Ninov clearly has an exceptionally brilliant and restless mind. Could his real passion have been, not to discover things about the natural world, but to pursue ideas, to solve intellectual challenges—the harder the better? And then I recalled that, as far as the superheavy elements were concerned, there *was* nothing to discover in the natural world. Those elements don't exist, they have to be created in the image of an idea about how matter should behave. So did Ninov, for all his labors on the detectors and the data-analysis software, see the challenge more in terms of philosophy than science? We may never know, but Walter Loveland holds out hope for an answer. "It would be interesting at some point in my life to sit down with Victor over a beer and talk candidly for a while," he said. "I don't know what I would hear."

Several weeks after my interview with Ninov, I received a message that shone an intriguing new light on the story. In the interview I had asked Ninov to suggest a scientist I might talk with who would support his point of view. He mentioned his former graduate advisor at GSI, Peter Armbruster, himself a renowned element hunter.

But Armbruster did not support Ninov's point of view. In an e-mail that he sent me in January 2007, he agreed with his German colleagues that Ninov fabricated two decay chains while he was at GSI. "I certainly feel deceived, and I can in no way justify what he has done," Armbruster wrote.

Perhaps more telling from a psychological perspective was this detail: Armbruster told me that Ninov's first fabricated decay chain occurred at 11:17 a.m. on November 11, 1995. November 11 marks the beginning of the south German carnival season known as Fasching, when people like to play all kinds of pranks. In Germany, the number 11 (*elf*) is known as *die närrische Zahl*—the fool's number. For this reason, the exact beginning of Fasching is *am elften elften elf Uhr elf,* which is to say "at 11:11 a.m. on November 11." Thus Ninov's first spurious decay chain occurred just five minutes after the official opening of Germany's practical-joke season.

"I suppose this chain . . . was composed by Victor Ninov as a joke for Fasching," wrote Armbruster. "Victor must have been very surprised that it was accepted and published. This success certainly encouraged Victor to go on, playing games with the group." Of course, there is no independent evidence that this was Ninov's motivation, and Ninov himself did not reply to an e-mailed request for a comment on Armbruster's theory.

A footnote: in 2006 the Dubna group, assisted by scientists from the Lawrence Livermore Laboratory (a separate institution from the Lawrence Berkeley Lab), announced that they had succeeded in creating three atoms of element 118. This they did by a completely different reaction from the one attempted at Berkeley. It was a hot-fusion reaction between calcium and californium, so the results said nothing about the validity of Smolańczuk's theory.

Both Gregorich and Loveland expressed some caution about the reported finding; they mentioned that most of the Russians' reported discoveries still await independent verification. But if the finding is verified, the Dubna group will get to name the new element. They may name it oganessium after their leader, Yuri Oganessian. They may name it after some historical Russian physicist—cherenkovium or zeldovium or even sakharovium. But they are unlikely to pick the name that seemed like a front-runner in 1999—ninovium.

Epilogue

There but for the grace of God go I. That is my own reaction to the stories just recounted, and I think most scientists would share it. There are so many opportunities for science to go wrong that scientists who reach the end of their careers without stumbling on one of them can count themselves not just smart or circumspect or morally superior, but also fortunate.

Of course I picked dramatic or memorable examples of scientific failure for this book. They're not typical of how science can go wrong, because mostly it does so in more mundane ways. Just as the spectacular successes of science are mere islands in a sea of worthy journeywork, so the spectacular failures are outnumbered by those that are slightly regrettable, modestly burdensome, or just partially incorrect. It is the destiny of most scientists to be neither canonized nor vilified by the judgment of history, but to be forgotten.

Similarly, most scientific accidents don't cause dozens of deaths, as the anthrax release at Sverdlovsk did. Most don't kill anyone, and of those that do, most kill just one person—the very person whose mistake caused the accident. One example: in August of 1996, Dartmouth College chemistry professor Karen Wetterhahn spilled a drop or two of dimethylmercury on her gloved hand. The drops penetrated both the glove and her skin, condemning her to a slow, painful death from mercury poisoning.

The example of scientific fraud recounted in this book—Victor Ninov's alleged fabrication of data supporting the discovery of a new chemical element—is also an extreme, unrepresentative case. Yes, there have been other cases that rival or outdo his: The fraudulent claim by South Korea's Hwang Woo-Suk to have created human stem cells by cloning is the most dramatic recent example. But most fraud consists of slight prettying-up or cherry-picking of data, omission of references to prior work in order to make one's own work seem more original, self-plagiarism, and the like.

Indeed, fraud merges imperceptibly into acceptable scientific practice. It's common, for example, for scientists to write up accounts of their research in which the sequence of experiments does not correspond to historical reality, or to introduce a paper with a hypothesis that wasn't actually formulated until after some of the experimental results came in. This is often thought to be justified: It aids comprehension to present the study as a logical sequence of ideas and experiments. But such deception causes harm if it leaves the reader thinking that a result was predicted by a hypothesis, or that a hypothesis was stimulated by prior results, when they were not. This can cause a scientist's conclusions to appear more believable than they actually warrant.

It would be an interesting exercise to go back in the scientific literature—say, twenty years or so—and pick a random selection of a hundred papers and ask, "Were they right in their main findings and conclusions, and were they as original as their authors claimed?" I don't know what fraction of them would have significant faults, but it would probably be substantial, and certainly much higher than most nonscientists would believe.

Most likely, those pieces of erroneous research would not have gone wrong in any memorable way—no conscious fraud, no switched labels, no blatant plagiarism—nor would they probably have any dire consequences. They probably resulted from countless trivial errors and omissions—the use of reagents whose specificity was less than expected, the selection of human subjects who were not fully representative of the group being investigated, the use of inappropriate statistical tests, or a lack of familiarity with the prior literature. Probably, in the ensuing decades, no one ever took the trouble to point out that the studies were wrong or to ask why; many scientific papers are not cited even once by other scientists, after all. The rising tide of scientific progress simply erases them from collective consciousness.

Still, science does sometimes go wrong in ways that are truly dramatic—accidents or drug trials in which people are injured or killed, erroneous claims that grab media attention and that take years to set right. And sometimes scientists themselves are appalled by the uses to which their discoveries are put by others. Take fetal ultrasound monitoring, a technique pioneered by the Scottish gynecologist and antiabortion campaigner Ian Donald, "My own personal fears are that my researches into early intrauterine life may yet be misused towards its more accurate destruction," wrote Donald in 1972. A decade or so later, ultrasound was being used to facilitate the abortion of millions of female fetuses in the third world.

Can anything be done that might cause science to go wrong less often? *Should* anything be done, even? These are thorny questions that are probably best left to professional ethicists or administrators or philosophers of science, but here are a few thoughts.

For a start, it's worth pointing out that it may take years or decades for the ill-effects of scientific discoveries and inventions to become evident. Take a field of applied science that I haven't covered in this book—industrial chemistry. In 1901 a German chemist, Wilhelm Normann, developed a process for turning vegetable oils into solid fats by hydrogenation. At the time, this invention seemed like an unalloyed benefit to humanity: It provided the means to produce inexpensive edible fats that resisted spoilage. It took more than half a century, and millions of premature deaths from heart disease, before the harmful effects of these fats on human health became apparent. In 1928, General Motors chemist Thomas Midgley, Jr., invented chlorofluorocarbon refrigerants—Freons. Again, the invention seemed to offer nothing but benefit to humanity, and it took decades before the downside—the destructive effect of these chemicals on the Earth's protective ozone layer—was understood. Looking back, it's hard to see how any program of regulatory oversight could have anticipated these dire consequences, given the lack of relevant knowledge at the time. In addition, there may never be agreement on the net benefit or harm of a discovery. It may depend on one's views about abortion, for example, in the case of Donald's invention. Thus preventing the long-term ill-effects of scientific inventions and discoveries is about as hard as predicting the future of civilization, and it is probably pointless to try.

Certainly, there can and should be oversight of science, especially in its applications. In medical research, for example, there are the Institutional Review Boards and federal regulatory bodies that do their best to see that research using human subjects is conducted ethically and safely. IRBs came up in several chapters of this book. I mentioned how their absence in the 1930s permitted unethical research such as Mary Tudor's stuttering study to go forward. I described how Robert Iacono circumvented IRBs and all other regulatory oversight by taking his patient to China for experimental surgery for Parkinson's disease. I also recounted, however, how Jesse Gelsinger died needlessly in a clinical trial that was overseen by a whole web of IRBs and government agencies.

It can be argued, however, that regulatory control and safety consciousness is a significant impediment to science—that it actually causes more harm than it averts. If we look back at some of the historical highlights of medical research, for example, we see what may look in hindsight like reckless risk-taking and a near-total lack of concern for ethical considerations.

William Harvey discovered the circulation of the blood through a series of experiments on unanesthetized animals that would turn the stomach of a modern reader. Edward Jenner picked an eight-year-old boy at random to test his first smallpox vaccine, then later inoculated him with smallpox in an effort to see if he had become immune to the disease—all, apparently, without so much as a by-your-leave. Walter Reed tested his mosquito theory for the transmission of yellow fever by having his colleagues expose themselves to insects that had fed on previous victims, causing one of his colleagues to develop a fatal infection. Thomas Starzl's first liver-transplant patient died shortly after surgery, as did every one of his patients over the next four years. Who would persist in the face of such odds: a madman—or a visionary?

To wish that none of these things had happened is to wish that none of those great advances saw the light of day. Risk-taking is part of scientific exploration, just as it was part of terrestrial exploration. No one expected that all those ships that set out for the Spice Islands would return safely home, and many didn't. Maybe we should allow the quest for the Magic Islands of the periodic table to take its victims too, even if they be self-inflicted victims like Victor Ninov.

This would be particularly true if there is a certain indivisible character trait that predisposes people both to the taking of great risks and to great scientific achievement. Many risk-taking scientists never make great discoveries, certainly, but few scientists make great discoveries without taking great risks—if only the risk of devoting a lifetime to the pursuit of a scientific will-o'-the-wisp. "My concern," James Wilson told the *New York Times* after Jesse Gelsinger's death, "is, I'm going to get timid, that I'll get risk averse." Which, in his mind at least, meant an end to productive science.

Of course, some of the episodes described in this book happened not in the process of scientific discovery, but during the application of scientific procedures to fairly mundane tasks—the identification of a rapist, the prediction of tomorrow's weather, the siting of a dam, or the production of a germ-warfare agent. In such cases, stricter oversight could hardly cramp scientific creativity. In fact as a result of disasters like the one that struck the St. Francis Dam, large engineering projects are now tightly regulated and reviewed, greatly reducing the risk of a repetition of a calamity. The Houston Crime Lab now operates under much closer oversight than was the case when Josiah Sutton was wrongly convicted. Weather forecasters are better trained and better equipped than they were before the Great October Storm. And germ-warfare agents, hopefully, are no longer being produced.

Even with the mishaps that involved genuinely scientific episodes, some might have been avoided by steps that did not impinge greatly on the process of science itself. Getting a volcanologist to wear a hardhat or to heed seismological warnings hardly seems like a major impediment to the study of volcanoes. Asking a neuroscientist to verify the identity of the drugs he is testing doesn't seem like putting a major roadblock in his path. Still, there may be an irreducible core of risk in science that cannot be eliminated without eliminating science's rewards. When one very successful scientist, tissue-engineering pioneer Robert Langer, was recently appointed to the Board of Directors of MIT's Whitehead Institute, the Institute's announcement included the following: "Bob Langer's work and life typifies so many of the strengths we aspire to at Whitehead—brash, audacious, risk-taking science." Somehow, Langer has parlayed that risk-taking trait into 800 scientific papers and many important discoveries, while at the same time avoiding all the traps that risk-taking makes scientists prone to. I sincerely hope that his science never does "go wrong" in any serious way, but if it should do so, I hope that he and others in his situation are judged for the entirety of their work, not for that one misstep. For it is so often just one such misstep—one momentary spasm of greed, haste, carelessness, credulity, or plain bad luck—that leads to disaster.

⚮

Sources

Chapter 1

Interviews:

Raymon Durso, M.D., July 5, 2000.
Rebecca Folkerth, M.D., July 5, 2000.
Robert Iacono, M.D., July 20, 2000.
Don L. Truex, D.D.S., September 2, 2005.
Mary Katherine (Kay) Truex De Justo, September 8, 2005.
James Slosson, 2005.

Scientific Publications:

R. P. Iacono, Z. S. Tang, J. C. Mazziotta, et al. (1992). "Bilateral fetal grafts for Parkinson's disease: 22 months' results." *Stereotactic and Functional Neurosurgery*, 58:84–87.

R. D. Folkerth and R. Durso (1996). "Survival and proliferation of nonneuronal tissues, with obstruction of cerebral ventricles, in a parkinsonian patient treated with fetal allografts." *Neurology*, 46:1219–1225.

J. H. Kordower, T. B. Freeman, R. A. E. Bakay, et al. (1997). "Treatment with fetal allografts." *Neurology*, 48:1737–1738.

News Accounts:

J. Perrone (1992): "Researchers circumvent fetal tissue transplant funding ban." *American Medical News*, June 8.

Other:

Division of Medical Quality, Medical Board of California: Decision and Order in the Matter of the Accusation against Robert Paul Iacono, M.D. September 12, 2005.

Chapter 2

Interviews:

Michael Fish, August 7, 2006
Bill Giles, August 14, 2006
Anita Hart, August 16, 2006
Thomas Jung, Ph.D., August 7, 2006
Ewen McCallum, M.Sc., August 17, 2006

Books:

Mark Davison and Ian Currie (1988). *Surrey in the Hurricane*. Froglets Publications.

Scientific Publications:

S. D. Burt and D. A. Mansfield (1988). "The great storm of 15–16 October 1987." *Weather (Royal Meteorological Society)* 43:90–114.
H. D. Lawes (1988). "The storm of 15–16 October 1987: A personal experience." *Weather (Royal Meteorological Society)* 43:142.
Meteorological Office (1988). "Report on the storm of 15/16 October 1987." *Meteorological Magazine* 117:97–140.
P. Swinnerton-Dyer and R. P. Pearce (1988). "Summary and conclusions from the Secretary of State's enquiry into the storm of 16 October 1987." *Meteorological Magazine* 117:141–144.
R. M. Morris and A. J. Gadd (1998). "Forecasting the storm of 15–16 October 1987." *Weather (Royal Meteorological Society)* 43:70–90.
T. Jung, E. Klinker, and S. Uppala (2004). "Reanalysis and reforecast of three major European storms using the ECMWF forecasting system. Part I: Analyses and deterministic forecasts." *Meteorological Applications* 11:243–261.

T. Jung, E. Klinker, and S. Uppala (2005). "Reanalysis and reforecast of three major European storms using the ECMWF forecasting system. Part II: Ensemble forecasts." *Meteorological Applications* 12:111–122.

Other:

Met Office: The great storm of 1987 (includes animation of the storm crossing the UK). Available at: www.met-office.gov.uk/education/secondary/students/1987.html.

BBC Weather: Michael Fish and the 1987 storm. Available at: www.bbc.co.uk/weather/bbcweather/forecasters/michael_fish_1987storm.shtml.

BBC Weather: Michael Fish MBE—The end of an era. (Includes video clip from Fish's forecast of October 15, 1987). Available at: www.bbc.co.uk/weather/bbcweather/forecasters/michael_fish_retirement.shtml.

University of Illinois at Urbana-Champaign: Online Meteorology Guide. Available at: ww2010.atmos.uiuc.edu/(Gh)/guides/mtr/home.rxml.

Worcester Heating Systems: Weathermen predict a warm front for Worcester. Available at: www.worcesterheatsystems.co.uk/index.php?fuseaction=site.newsDetail&con_id=5258&detail_id=128532.

Chapter 3

Interviews:

Bernard Chouet, May 23, 2006.
Charles Wood, May 12, 2006.

Books:

Victoria Bruce (2001). *No Apparent Danger: The True Story of Volcanic Disaster at Galeras and Nevado del Ruiz*. HarperCollins.

Stanley Williams (2001). *Surviving Galeras*. Houghton Mifflin.

Scientific Publications:

J. Stix, M. L. Calvache, and S. N. Williams (eds.) (1997). "Galeras Volcano, Colombia: Interdisciplinary study of a Decade Volcano." *Journal of Volcanology and Geothermal Research* 77:1–338.

H. Sigurdsson (2001). "Volcanology: Not a 'piece of cake.'" *Science* 292:643–644.

News account:

R. Monastersky (2001). "Under the volcano: Deaths rattle one of the riskiest disciplines in science." *Chronicle of Higher Education*, March 30. Available at: www.chronicle.com/free/v47/i29/29a01801.htm.

International Association of Volcanology and Chemistry of the Earth's Interior (1994). Recommended safety measures for volcanologists. Available at: www.volcanology.geol.ucsb.edu/safety.htm.

Chapter 4

Interviews:

Colin Blakemore, Ph.D., June 19, 2006
Rick Doblin, Ph.D., June 14/15, 2006
Charles Grob, M.D., June 16, 2006
George Ricaurte, M.D., Ph.D., July 26, 2006

Scientific Publications:

U. D. McCann, Z. Szabo, et al. (1998). "Positron emission tomographic evidence of toxic effect of MDMA ('Ecstasy') on brain serotonin neurons in human beings." *Lancet* 352:1433–7.

C. S. Grob (2000). "Deconstructing Ecstasy: The politics of MDMA research." *Addiction Research* 8:549–588.

G. A. Ricaurte, J. Yuan, et al. (2002). "Severe dopaminergic neurotoxicity in primates after a common recreational dose regimen of MDMA ('ecstasy')." *Science* 297:2260–3.

G. A. Ricaurte, J. Yuan, et al. (2002). Response. *Science* 300:1504–1505.

G. A. Ricaurte, J. Yuan, et al. (2003). Retraction. *Science* 301:1479.

M. Mithoefer, L. Jerome, and R. Doblin (2003). Letter: MDMA ("Ecstasy") and neurotoxicity. *Science* 300:1504–1505.

U. D. McCann, Z. Szabo, et al. (2005). "Quantitative PET studies of the serotonin transporter in MDMA users and controls using [11C]McN5652 and [11C]DASB." *Neuropsychopharmacology* 30:1741–50.

G. A. Ricaurte, A. O. Mechan, et al. (2005). "Amphetamine treatment similar to that used in the treatment of adult attention-deficit/hyperactivity disorder damages dopaminergic nerve endings in the striatum of adult nonhuman primates." *Journal of Pharmacology and Experimental Therapeutics* 315:91–8.

R. Buchert, R. Thomasius, et al. (2006). "Reversibility of ecstasy-induced reduction in serotonin transporter availability in polydrug ecstasy users." *European Journal of Nuclear Medicine and Molecular Imaging* 33:188–99.

News Accounts:

D. G. McNeil, Jr. (2003). "Research on Ecstasy Is Clouded by Errors." *New York Times*, December 2.

R. Walgate (2002). Independent inquiry demanded into Ecstasy affair. *The Scientist*, September 18.

R. Highfield, (2002). "Ecstasy Use 'May Lead to Parkinson's-type Illness.'" *The Telegraph* (London), September 27.

Other:

Charles Grob: Testimony before the U.S. Sentencing Commission on MDMA. Available at: at www.maps.org/news/grobtestimony.html.

Congressional Record, June 18, 2002: Reducing Americans' Vulnerability to Ecstasy Act of 2002. Available at: frwebgate.access.gpo.gov/cgi-bin/useftp.cgi?IPaddress=162.140.64.31&filename=s2633is.txt&directory=/disk2/wais/data/107_cong_bills

Drug Enforcement Agency (2002). Statement of Asa Hutchinson before the House Judiciary Subcommittee on Crime, Terrorism and

Homeland Security. Available at: www.usdoj.gov/dea/pubs/cngr
test/ct101002.html.

National Institute on Drug Abuse: Director's Report to the National
Advisory Council on Drug Abuse—February, 2003. Available at:
www.drugabuse.gov/DirReports/DirRep203/DirectorReport1
.html.

Emergency Department Trends from the Drug Abuse Warning Network, 1994–2001. Available at: www.oas.samhsa.gov/DAWN/
Final2k1Edtrends/text/Edtrend2001v6.pdf.

CNN Student News (2000). Famous fried eggs: Students debate effectiveness, accuracy of well-known anti-drug commercial. Available
at: www.cnnstudentnews.cnn.com/fyi/interactive/news/brain/brain
.on.drugs.html.

Multidisciplinary Association for Psychedelic Studies (2004). NIDA
and NCRR funding for Ricaurte and McCann 1989–2003. Available at: www.maps.org/mdma/ricaurtefunding.pdf.

Multidisciplinary Association for Psychedelic Studies (undated).
NIDA Grant Progress Report, George A. Ricaurte, Principal Investigator, July 17, 2003, (obtained by an FOIA request). Available at:
adgjm.net/ura/ua.cgi?ul=http%3A%2F%2Fwww.maps.org%2Fm
dma%2Fretraction%2Ffoia012804%2Fprogress1.html&ua=

Chapter 5

Interviews:

Jack Green, Ph.D., March 1, 2006.

Books:

Charles F. Outland (1963, revised 2002). *Man-Made Disaster: The
Story of St. Francis Dam.* Historical Society of Southern California
and Vroman's Bookstore.

Doyce B. Nunis., ed. (2002). *The St. Francis Dam Disaster Revisited.* Historical Society of Southern California and Vroman's
Bookstore.

Scientific Publications:

J. D. Rogers (2002). "A man, a dam, and a disaster: Mulholland and the St. Francis Dam." In Nunis (see page 278), pp. 1–109.

Other:

A. J. Wiley, G. D. Louderback, F. L. Ransome, et al. (1928). *Report of the Commission Appointed by Governor C.C. Young to Investigate the Causes Leading to the Failure of the St. Francis Dam near Saugus, California.* State of California.

J. D. Rogers. Failure of the St. Francis Dam. Available at: web.umr .edu/~rogersda/st_francis_dam.

Jack Green. St. Francis Dam. Available at: www.seis.natsci.csulb. edu/VIRTUAL_FIELD/Francesquito_Dam/franmayn.htm.

Goutal, N. The Malpasset Dam failure: An overview and test case definition. Available at: www.hrwallingford.co.uk/projects/CADAM/ CADAM/zaragoza/z2.pdf.

Chapter 6

Interviews:

Arthur Caplan, Ph.D., November 24, 2006.

Paul Gelsinger, November 27, 2006 (and subsequent e-mail correspondence).

Robert Erickson, M.D., December 6, 2006.

Inder Verma, Ph.D., December 4, 2006.

Scientific Publications:

X. Ye, M. B. Robinson, M. L. Batshaw, et al. (1996). Prolonged metabolic correction in adult ornithine transcarbamylase-deficient mice with adenoviral vectors. *Journal of Biological Chemistry,* 271:3639–46.

National Institutes of Health Recombinant DNA Advisory Committee (2002). Assessment of adenoviral vector safety and toxicity. *Human Gene Therapy,* 13:3–13.

D. A. Sisti and A. L. Caplan (2003). Back to basics: Gene therapy research ethics and oversight in the post-Gelsinger era. In *Ethik und Gentherapie: Zum praktischen Diskurs um die molekulare Medizin* (C. Rehmann-Sutter and H. Müller eds.), Francke Verlag.

G. Suntharalingam, M. R. Perry, S. Ward, et al. (2006). Cytokine storm in a phase-1 trial of the anti-CD28 monoclonal antibody TGN1412. *New England Journal of Medicine* 355:1018–1028.

News Accounts:

S. G. Stolberg (1999). "The Biotech Death of Jesse Gelsinger." *New York Times* (November 28).

S. G. Stolberg (1999). "FDA Officials Fault Penn Team in Gene Therapy Death." *New York Times,* December 9.

D. Nelson and R. Weiss (1999). "Hasty Decisions in the Race to a Cure? Gene Therapy Study Proceeded Despite Safety, Ethics Concerns." *Washington Post,* November 21.

E. Marshall (2000). "Gene Therapy on Trial." *Science* 288:951–957.

D. Sanderson (2006). "I Was Treated No Better Than an Animal, Says Drug victim." *The Times (London),* March 31.

D. Leppard (2006). "Elephant Man Drug Victims Told to Expect Early Death." *Sunday Times (London),* July 30.

E. Rosenthal (2006). "Ill-fated UK Drug Trial Bares Testing Loopholes." *International Herald Tribune* (July 30).

Other:

FDA: Letter to James Wilson dated March 3, 2000: www.fda.gov/foi/warning_letters/m3435n.pdf.

FDA: Letter to James Wilson dated February 8, 2002: www.fda.gov/foi/nooh/Wilson.htm.

FDA: Letter to Mark Batshaw dated November 30, 2000: www.fda.gov/foi/nidpoe/n14l.pdf.

Institute for Human Gene Therapy (2000): Institute for Human Gene Therapy Responds to FDA (press release, February 14). www.upenn.edu/almanac/between/FDAresponse.html.

United States Attorney's Office, Eastern District of Pennsylvania (2006). US settles case of gene therapy study that ended with teen's death (press release, February 9). Available at: www.usdoj.gov/usao/pae/News/Pr/2005/feb/UofPSettlement%20release.html.

P. Gelsinger, (2001). *Jesse's intent.* www.circare.org/jintent.pdf.

Independent panel reviewing the University of Pennsylvania's Institute for Human Gene Therapy (2000). *Report.* Available at: www.upenn.edu/almanac/v46/n34/IHGT-review.html.

J. M. Wilson (1999). Presidential address, American Society of Gene Therapy. web.archive.org/web/20001210180800/www.med.upenn.edu/~ihgt/info/asgt99.html.

Chapter 7

Interviews:

Stephen Hanauer, 2005 (date unknown).

Books:

William McKeown (2003). *Idaho Falls: The Untold Story of America's First Nuclear Accident.* ECW Press.

Susan M. Stacy (2000). *Proving the Principle: A History of the Idaho National Engineering and Environmental Laboratory, 1949–1999.* U.S. Government Printing Office.

Other:

Arlington National Cemetery Web site: Richard Leroy McKinley. www.arlingtoncemetery.net/rlmckinl.htm.

U.S. Atomic Energy Commission, Idaho Operations Office. *The SL-1 Accident.* Distributed by Radiationworks.com (documentary film).

Chapter 8

Interviews:

Ken Alibek, April 13, 2006.

Jeanne Guillemin, April 2006 (e-mail correspondence).

Peter Gumbel, April/May 2006 (e-mail correspondence).
Matthew Meselson, April 17, 2006.

Books:

Ken Alibek (1999). *Biohazard: The Chilling True Story of the Largest Covert Biological Weapons Program in the World—Told from the Inside by the Man Who Ran It.* Random House.

Jeanne Guillemin (1999). *Anthrax: The Investigation of a Deadly Outbreak.* University of California Press.

Tom Mangold and Jeff Goldberg (1999). *Plague Wars: The Terrifying Reality of Biological Warfare.* St. Martin's Press.

Judith Miller, Stephen Engelberg, and William Broad (2001). *Germs: Biological Weapons and America's Secret War.* Simon & Schuster.

Scientific Publications:

M. Meselson, J. Guillemin, J. Martin Hugh-Jones, et al. (1994). The Sverdlovsk anthrax outbreak of 1979. *Science*, 266:1202–1208.

J. B. Tucker, The "Yellow Rain" controversy. Lessons for arms control compliance. *The Nonproliferation Review*, Spring 2001, pp. 25–42.

J. Guillemin (2002). The 1979 anthrax epidemic in the USSR: Applied science and political controversy. *Proceedings of the American Philosophical Society*, 146:18–36. Available at: www.aps-pub.com/proceedings/1461/102.pdf.

D. A. Wilkening (2006). Sverdlovsk revisited: Modeling human inhalation anthrax. *Proceedings of the National Academy of Sciences*, 103:7589–7594.

Other:

Robert A. Wampler and Thomas S. Blanton (eds.). The National Security Archive—Volume V: Anthrax at Sverdlovsk, 1979. Available at: www.gwu.edu/~nsarchiv/NSAEBB/NSAEBB61/.

Chapter 9

Interviews:

William Thompson, J.D., Ph.D., July 7, 2006.

Scientific Publications:

W. C. Thompson, F. Taroni, and C. G. G. Aitken (2003). How the probability of a false positive affects the value of DNA evidence. *Journal of Forensic Science* 48:1–8.

News Accounts:

M. A. Fergus, (2004). "Josiah Sutton: One Year Later." *Houston Chronicle*, March 6.

R. Khanna, (2006). "In a Lot of Ways, I Still Am Suffering." *Houston Chronicle*, June 23.

M. Possley (2006). "Inmate Wrongly Executed." *Chicago Tribune*, May 3.

Associated Press (2006). "Experts Question Arson Convictions," May 7. Available at: spd.iowa.gov/filemgmt_data/files/arson.pdf.

R. Khanna and S. McVicker (2004). "Fired DNA Analyst to Return to Work at Crime Lab." *Houston Chronicle*, January 27.

Other:

Houston Chronicle: Hot topic: HPD Crime Lab—Index of stories. Available at: www.chron.com/content/chronicle/special/03/crimelab/index.html.

W. C. Thompson, (2003). Review of DNA evidence in *State v. Josiah Sutton*. Available at: www.scientific.org/archive/Thompson%20Report.pdf.

Home page, Office of the Independent Investigator for the Houston Police Department Crime Laboratory and Property Room [Michael Bromwich's reports]. Available at: www.hpdlabinvestigation.org/.

National Research Council. The evaluation of forensic DNA evidence: An update. Available at: www.nap.edu/readingroom/books/DNA.

Chapter 10

Interviews:

John Casani, Ph.D., July 29, 2006.
Steve Jolly, Ph.D., September 12, 2006.
Sam Thurman, Ph.D., August 31, 2006.

News Accounts:

J. Kaye (2000): "NASA in Question." *PBS NewsHour*, April 14. Available at: www.pbs.org/newshour/bb/science/jan-june00/nasa_4-14.html.

M. Munro (2004): "Ryan: Death of Four-Year-Old Is a Tragic Reminder of the Dangers of Clinical Research." *CanWest News Service*, February 26. Available at: www.michenerawards.ca/english/ryan.htm.

Other:

NASA-TV: Mars Climate Orbiter Orbit Insertion Event, September 23, 1999. Courtesy of JPL AudioVisual Services Office.

Mars Climate Orbiter Mishap Investigation Board: Phase I Report. Available at: ftp.hq.nasa.gov/pub/pao/reports/1999/MCO_report.pdf.

Mars Climate Orbiter Mishap Investigation Board: Report on Project Management in NASA. Available at: ftp.hq.nasa.gov/pub/pao/reports/2000/MCO_MIB_Report.pdf.

NASA-JPL: Basics of space flight, chapter 13: space navigation. Available at: www2.jpl.nasa.gov/basics/bsf13-1.html.

Food and Drug Administration: Warning letter to Dr. Jacqueline Halton, April 14, 2003. Available at: www.fda.gov/foi/warning_letters/g3946d.htm.

Chapter 11

Interviews:

Nicoline Ambrose, Ph.D., October 17, 2006.
Oliver Bloodstein, Ph.D., October 9, 2006 (e-mail).

Curtis Krull, J.D., October 24, 2006.
Richard Schwartz, Ph.D., October 16, 2006.

Books:

Wendell Johnson (1930). *Because I Stutter*. D. Appleton and Co. Available at: www.uiowa.edu/~cyberlaw/wj/bis/wjbis.html.
Wendell Johnson (1961). *Stuttering and What You Can Do About It*. University of Minnesota Press. Available at: www.uiowa.edu/~cyberlaw/wj/wjswycda.html.

Scientific Publications:

W. Johnson (1938). The role of evaluation in stuttering behavior. *Journal of Speech Disorders* 3:85–89.
M. Tudor (1939). *An experimental study of the effect of evaluative labeling on speech fluency*. M.A. thesis, Department of Psychology, University of Iowa.
N. G. Ambrose and E. Yairi (2002). The Tudor study: Data and ethics. *American Journal of Speech-Language Pathology* 11:190–203.
E. Yairi (2005). The Tudor study and Wendell Johnson. In *Ethics: A Case Study from Fluency*. Plural Publishing.
R. G. Schwartz (2005). Would today's IRB approve the Tudor study? In *Ethics: A Case Study from Fluency*. Plural Publishing.
N. Johnson (2005). Retroactive ethical judgments and human subjects research: The 1930 Tudor study. In *Ethics: A case study from fluency*. Plural Publishing. Available at: www.nicholasjohnson.org/wjohnson/hsr/njhsr512.pdf.

News Accounts:

J. Dyer (2001). "Ethics and Orphans: The 'Monster Study.'" *San Jose Mercury News,* June 10.
J. Dyer (2001). "Theory Improved Treatment and Understanding of Stuttering." *San Jose Mercury News,* June 11.
C. Krantz (2001). "Reporter Quits in Stutterers Case." *Des Moines Register,* August 1.

G. Reynolds (2003). "The Stuttering Doctor's 'Monster Study.'" *New York Times,* March 16.

Chapter 12

Interviews:

Peter Armbruster, Ph.D., January 8, 2007 (e-mail).
Kenneth Gregorich, Ph.D., December 11, 2006.
Walter Loveland, Ph.D., December 14, 2006.
Victor Ninov, Ph.D., December 14, 2006.

Books:

Darleane C. Hoffman, Albert Ghiorso, and Glenn T. Seaborg (2000). *The Transuranium People: The Inside Story.* Imperial College Press.

Scientific Publications:

V. Ninov, K. E. Gregorich, W. Loveland, A. Ghiorso, D. C. Hoffman, et al. (1999). Observation of superheavy nuclei produced in the reaction of ^{86}Kr with ^{208}Pb. *Physical Review Letters* 83:1104–1107.

S. Hofmann, F. P. Hessberger, D. Ackermann, et al. (2002). New results on elements 111 and 112. *European Physical Journal A* 14:147–157.

Editorial note [retraction] (2002): Observation of superheavy nuclei produced in the reaction of ^{86}Kr with ^{208}Pb. *Physical Review Letters* 89:9901 (15 July).

Y. T. Oganessian, V. K. Utyonkov, Y. V. Lobanov, et al. (2006): Synthesis of the isotopes of the elements 118 and 116 in the ^{249}Cf and ^{245}Cm+^{48}Ca reactions. *Physics Review* C 74:e044602.

News Accounts:

Anonymous (1999): "Discovery of New Elements Makes Front Page News." *Berkeley Lab Research Review* (Summer). Available at:

www.lbl.gov/Science-Articles/Research-Review/Magazine/1999/departments/breaking_news.shtml.

G. Johnson (2002): "At Lawrence Berkeley, Physicists Say a Colleague Took Them for a Ride." *New York Times,* October 15.

K. Davidson (2002): "Accusations of Fraud: Fired Lawrence Lab Physicist's Earlier Data Also Questioned." *San Francisco Chronicle,* July 21.

Other:

M. Gilchriese, A. Sessier, G. Trilling, and R. Vogt (2002). *Report of the committee on the formal investigation of alleged scientific misconduct by LBNL staff scientist Dr. Victor Ninov.* Lawrence Berkeley National Laboratory.